PERIPHERAL NEUROPATHY

Edited by **Paulo Armada-da-Silva**

Peripheral Neuropathy
http://dx.doi.org/10.5772/57082
Edited by Paulo Armada-da-Silva

Contributors

James L. Henry, Haigang Gu, Zhilian Yue, Qi Zhang, Paulo Armada-Da-Silva, Cátia Pereira, Sandra Amado, Ana Luis, Rahul Nath, Chandra Somasundaram, Takashi Kawano

Published by InTech

Janeza Trdine 9, 51000 Rijeka, Croatia

Publishing Process Manager Iva Lipovic

Technical Editor InTech DTP team

Cover InTech Design team

First published May, 2014

Printed in Croatia

Additional hard copies can be obtained from orders@intechopen.com

Peripheral Neuropathy, Edited by Paulo Armada-da-Silva

p. cm.

ISBN 978-953-51-1451-2

For EU product safety concerns:
IN TECH d.o.o., Prolaz Marije Krucifikse Kozulić 3, 51000 Rijeka, Croatia,
info@intechopen.com or visit our website at intechopen.com.

Meet the editor

Paulo Armada-Da-Silva is a lecturer of Anatomy and Physiology in the Department of Sports and Health, Faculty of Human Kinetics, University of Lisbon, and a former staff from the Physiotherapy Department in Lisbon central hospital, where he gained experience in the rehabilitation of central and peripheral nerve injury patients. He is a current member of the Neuromechanics of Human Movement research group with a special interest in neurorehabilitation and in the role of physical activity on regeneration of the peripheral nerves. He collaborates with other groups working in peripheral nerve research and regenerative medicine, which share an interest in developing novel solutions for the reconstruction of damaged nerves and how they might be combined with rehabilitation, in order to develop a comprehensive approach for the treatment of peripheral nerve damage.

Contents

Preface

Peripheral nerve dysfunction occurs as a result of many different events, including nerve injury caused by trauma or by other mechanisms, diseases of neurological or metabolic nature, and infections, among other reasons. In parallel with different causes, peripheral neuropathy is also characterized by plentiful variation in the type, number and anatomical location of the nerves that are affected. Hence, peripheral nerve dysfunctions are typically complex, either because they affect different nerves or because they cause multiple disturbances of motor, sensory and autonomic functions. As a consequence of the complex nature of peripheral nerve dysfunctions, their management usually requires the contribution of health professionals with different competences and skills. In few cases, peripheral nerve dysfunctions resolve spontaneously, without the need for further intervention, such as in the case of minor traumas or infections. Most frequently, however, the peripheral nerve damage is severe and extensive denervation ensues. In these unfortunate situations, the recovery of motor and sensory functions in the affected body region occurs only partially or does not occur at all.

Together with motor and sensory sequels, peripheral nerve dysfunction is also associated with positive signs and symptoms. A significant number of patients suffering from peripheral nerve damage develop neuropathic pain. This is usually very disabling and neuropathic pain is often referred to by patients as a prominent reason for impoverished quality of life. As for other consequences of peripheral nerve damage, the mechanisms triggering neuropathic pain may reside on injury signaling but are also probably related with the complex array of molecular and cellular responses set in motion by the injury. Many of such responses have now been described as a result of the research effort conducted over the last decades. Hence, crucial neurobiological factors determining the ability of the peripheral nerves to regenerate and reinnervate their targets have been disclosed. These include the decisive role played by nerve supportive cells, in particular Schwann cells, and the importance of the scaffold provided by the regenerating nerve itself. In this respect, the timing of treatment stands as a key factor for nerve regeneration and end-organ reinnervation success, as the regenerative potential of the nerve decays with time.

The goal of this book is to give an account of current approaches in the field of peripheral neuropathy along different perspectives. This book combines research-driven information and more clinically-oriented topics, bearing on firm clinical experience. Approaching peripheral neuropathy from different, although complementary, points of view, the book incorporates a number of chapters that were organized according to their emphasis. The chapters should offer the reader both a general idea about the purpose and achievements of current research and more personal points of view, driven by clinical experience and clinical evidence.

Chapter 1 by Dr. James opens the book and addresses the importance of paying more attention to the heightened risk of peripheral nerve damage nowadays. The chapter analyses why peripheral nerve injuries are becoming more common in modern societies and drives the attention to the cost of these injuries, both in terms of health care costs and in the quality of life of those affected by such events.

Chapter 2 by Dr. Haigang and co-workers addresses the importance of tissue engineering in treating damaged peripheral nerves. Guiding tubes are already in use for the reconstruction of small nerve defects, but the continued development of newer biomaterials and the application of cell therapy will hopefully allow their use in the treatment of larger nerve defects in the future. This chapter specifically analyses the potential role of stem cells-derived neural cells and biodegradable polymers in the construction of viable guiding tubes that could then be used as substitutes for nerve autografts.

Enhancing nerve regeneration, reinnervation and functional recovery following peripheral nerve injury by physical activity and activity-based strategies is addressed in our own Chapter 3. This chapter offers an in-depth review of research demonstrating the role of physical exercise on promoting axonal growth. It also provides a description of molecular and cellular mechanisms that possibly underlie the action of activity-based interventions. Also, briefly described is the important role played by the plasticity and modulation of spinal and supraspinal neural circuitry driven by activity-based interventions in improving functional recovery.

Chapter 4 by Dr. Nath and Dr. Somasundaram describes surgical treatment of obstetric brachial plexus palsy, supported on extensive clinical experience and successful outcomes. By focusing on correcting skeletal deformities, the surgical approach presented in this chapter drives our attention to the close relationship between innervation, movement and the skeleton and how intervening at such distinct levels is important to enabling function.

Chapter 5 by Dr. Kawano examines diabetic peripheral neuropathy and describes its mechanisms, as well as the procedures for diagnosing and treating this condition.

We wish that by reading this book the reader will be offered a general overview of current research and clinical practice surrounding peripheral neuropathy and the many venues for progress in this field.

Paulo Armada-da-Silva, Ana Colette Maurício and Stefano Geuna
Department of Sports and Health,
Faculdade de Motricidade Humana,
Universidade de Lisboa, Cruz Quebrada,
Portugal

Chapter 1

Mechanisms of Peripheral Nerve Injury – What to Treat, When to Treat

James L. Henry

Additional information is available at the end of the chapter

http://dx.doi.org/10.5772/58421

1. Introduction

Peripheral nerve injury, sometimes referred to as acquired nerve injury, is a catastrophic injury that imposes a number of negative outcomes that usually inflict one or more adverse health conditions or disabilities on its victims. These adverse health conditions and disabilities frequently place both short-term and long-term burdens on individuals, families, communities, the workplace, the health care system and economies in general. Considerable money and effort have been expended on attempts to lessen, prevent or ameliorate the effects of trauma on peripheral nerves.

There are many outcomes of neuropathy of peripheral nerves. Neuropathic pain is perhaps the best documented, largely because of the enormous impact of chronic neuropathic pain on individuals and the fact that it tends to be refractory to medical treatment [1, 2]. However, other outcomes of secondary injury to peripheral nerves include, in terms of sensory disturbance, numbness, dysesthesia (an unpleasant abnormal sensation, whether spontaneous or evoked), paresthesia (an abnormal sensation, such as tingling, whether spontaneous or evoked), hypoesthesia (decreased sensitivity to stimulation, excluding the special senses) and loss of proprioception (possibly contributing to altered gait and to falls). In terms of motor control, peripheral neuropathy can lead to weakness, loss of movement, loss of corrective motor control and loss of muscle mass. Neuropathy of the autonomic nervous system can be manifest as orthostatic hypotension, dysautonomia, altered sudomotor function, and the like. Injury or damage to nerves or nerve cells can be the result of factors or events that are unanticipated or unexpected, such as an accident, while other factors or events can be anticipated, such as a result of chemotherapy or even surgical intervention.

It is the thesis of the present overview that many of the negative outcomes and disabilities of peripheral nerve injury can be reduced in severity, or prevented altogether, by early intervention with the appropriate methods, procedures and pharmaceutical formulations, continued for a medically-beneficial period of time.

While there are currently practices and interventions to treat, manage or diminish the negative outcomes of peripheral nerve injury once they have been established, immediate or early preventive approaches targeted at the development phase of these outcomes are nonexistent, few or ineffective. That is, at the time of a traumatic event medical attention focuses on treatment of immediate symptoms such as bleeding or to avoid infection, but medical attention does not typically address treatment to prevent the cascade of restorative, or adaptive, and degenerative, or maladaptive, metabolic and biochemical processes that result from peripheral nerve injury and that lead to prolonged or permanent adverse health conditions and disability.

Peripheral nerve injury or damage is not only to nerves or nerve cells, but can include damage to neural support cells, such as satellite cells and myelin cells, and also local circulation. In the context of the present overview the term "neural support cell" is any cell that supports or could be considered to support the health, normal function and survival of nerves and nerve cells, and include myelin cells and satellite cells, astroglia, oligodendrocytes, Schwann cells, vascular endothelial cells, and the like. Further, the term "neural support tissue" is any tissue that supports or could be considered to support the health, normal function, phenotype, gene expression or survival of nerves, nerve cells or support cells, and include the vasculature or microvasculature, particularly the endothelial cells that prevent blood from leaking into peripheral nerve tissue and that provide the selective blood-nerve barrier that allows the passage of certain supportive chemicals into nerve tissue as well as the passage of nerve tissue wastes out of nerve tissue.

Degeneration of axons or of neural support cells triggers a cascade of activated chemical pathways that lead to injury to otherwise intact nerve cells, Schwann cells, local vasculature and even more remote nerve cells by entry of degradation products into the blood circulation.

Trauma can be of many different types. For example, physical trauma can occur in vehicle accidents, in workplace accidents, sports accidents, on the battlefield, from falls, from assaults, from landmines and from explosive or other blasts, and the like, but can also occur as a result of surgical or other medical procedures.

Chemical trauma to nerves or nerve cells or neural support cells or neural support tissues can occur, from alcohol overdose, drug abuse, stimulant drugs such as pentylenetetrazol, carbon dioxide poisoning, acrylamide and related chemicals, overexposure to certain environmental chemicals such as copper or natural hazards such as scorpion venom toxin, herbicides, agricultural insecticides such as lindane, many industrial chemicals, neurotoxin bioterrorism chemicals such as soman and sarin, as well as radiation bioterrorism chemicals such as polonium and strontium.

Medically-induced trauma to nerves or nerve cells or neural support cells or neural support tissues can occur as a result of surgery, amputation, injections, laparoscopy, implants, during a medical procedure that reduces or impedes the blood supply to any tissue containing nerve cells for any period of time as described herein, chemotherapy (for example from methotrexate, cisplatin, cytosine arabinose, carmustine, thiotepa among others), radiation therapy, immunosuppressants (such as tacrolimus).

Surgery in and of itself can produce other types of trauma that are injurious to nerves, nerve cells and nerve support cells and neural support tissues. For example, the life-saving benefits of cardiac surgery are frequently followed by manifestations of damage to peripheral nerves and chronic neuropathic pain, which remains a significant complication of cardiac surgery and occurs with an incidence of 15% [3]. In a ten-year overview it was reported that chronic post-surgical pain constitutes a significant medical need that may be amenable to be reduced or prevented [4]. The incidence of persisting pain following amputation is 50-85%, that of intrathoracic surgery is 30-55%, mastectomy is 20-50%. Persisting pain after even simple procedures such as hernia repair and cholecystectomy are reported to be 5-35% and 5-50%, respectively. Post-sternectomy pain has been reported to be as high as 28% [5], hysterectomy as high as 32% [6], caesarean section as high as 12-18% [7, 8]. These are all astounding numbers, especially given the position taken here that the persisting pain may be reduced or prevented by appropriate and timely pharmaceutical intervention.

Similarly to medical surgical procedures, the incidence of peripheral nerve injury resulting from medical chemotherapy procedures has been underappreciated. A recent systematic review [9] has calculated that the prevalence of patients with neuropathic pain as a result of chemotherapy varies between 19% and 39%. Neuropathology-inducing chemotherapy drugs include paclitaxel, bortezomib, ixabepiline and oxaliplatin, as examples.

Broader than chemotherapy, peripheral neuropathy resulting from medications and toxic chemicals may also be preventable. Drugs that can trigger peripheral neuropathy include some antibiotics, as well as drugs for other disorders, including infliximab, etanercept, leflunomide, linezolid, statins, dichloroacetate and others [10-12]. A number of industrial chemicals also have the potential to induce peripheral neuropathy [12, 13]. Based on available knowledge, these types of peripheral neuropathy can be reduced or prevented if the appropriate medical intervention is applied early.

2. Secondary injury resulting from peripheral nerve trauma

Peripheral nerve injury generally consists of two related processes, the initial injury and a secondary injury [14] that results from cascades of self-propagating metabolic and biochemical processes that lead to loss of cell function and cell death. Disability occurs to a large extent from the secondary injury that is triggered by the primary injury. Whether the numbers are large, as in the case of battlefield peripheral nerve injury, or small, as with falls in the elderly, peripheral nerve injury can be devastating to the individual. Whether the numbers are large, as those resulting from car accidents, or small, such as those resulting from laparoscopic surgery, the result of peripheral nerve injury can be a future of constant burning, debilitating neuropathic pain and other adverse health conditions.

In the context of the present overview the term "secondary injury" or "secondary damage", terms that can be used interchangeably for the present context, means any damage, injury, harm, loss, change in structure, change in phenotype, change in gene expression or change in function or survival of nerves, nerve cells, neural support cells or neural support tissue that occurs after a traumatic event and develops over the seconds, minutes, hours, days, weeks or even months following such an event. Secondary injury or secondary damage is usually considered to result from consecutive or parallel biochemical cascades of cellular and metabolic processes that are activated or triggered by the trauma-induced direct tissue damage to a peripheral nerve. Secondary injury or secondary damage is usually considered to involve endogenous processes or biosynthetic pathways that govern, regulate or influence the structure, health, function, gene expression or survival of nerves or nerve cells, or cells or tissues upon which nerves or nerve cells depend to maintain health and function, such as neural support cells and neural support tissues often with delayed clinical presentation [14]. Whereas a primary injury is irreversible [15], this secondary injury is salvageable [14]. This secondary injury is a neuropathology that can be reduced or prevented by appropriate and timely intervention.

3. Functional outcomes of peripheral nerve injury

Neuropathic pain is defined as pain caused by damage, lesion, disease or altered function of the peripheral somatosensory nervous system and is characterized as a constant burning pain accompanied by hyperesthesia (increased sensitivity to stimulation, excluding the special senses). Hyperesthesia is usually clarified in clinical use as either hyperalgesia (increased pain from a stimulus that normally provokes pain) or allodynia (pain due to a stimulus that does not normally provoke pain). Neuropathic pain may also include periodic attacks of pain that feel like electric shocks or shooting pain.

In understanding mechanisms of peripheral neuropathic pain it is important to distinguish this type of pain from nociceptive pain and inflammatory pain. Nociceptive pain is pain that arises from actual or potential damage to non-neural tissue and is due to the activation of nociceptors in these tissues. Nociceptive pain functions to protect from potential tissue damage or from further tissue damage, and triggers physiological and behavioural reflexes to avert damage. Inflammatory pain is pain that is triggered by inflammation and serves to aid in healing and repair of injured non-neural tissue. Each of nociceptive and inflammatory pain is brought about by a specific set of mechanisms and each has a specific treatment algorithm. Neuropathic pain is different, not only with respect to underlying mechanisms and treatments [16], but it is considered to be a maladaptive pain, as this type of pain neither protects nor supports healing and repair.

4. Processes and mechanisms leading to peripheral neuropathology following peripheral nerve trauma

Trauma to a peripheral nerve has effects on sensory neurons, on motor neurons controlling skeletal muscle and on autonomic efferent neurons controlling the cardiovascular system and organs. This accounts for the range of outcomes of peripheral nerve trauma indicated above. As much of the research in this area is focussed on symptoms, available knowledge tends to be clearly fractionated into the bases of sensory loss, motor loss or autonomic dysfunction.

The overriding research on sensory loss pertains to peripheral neuropathic pain, which will be surveyed here to exemplify the pathophysiological changes that occur in peripheral nerves more generally. Even here, though, much attention is focused on changes in the central nervous system [17], particularly at the level of the first sensory synapse in the spinal cord and in the brain stem [18, 19]. As a result, treatment options tend to focus on targets within the central nervous system [20-22]. Research also tends to focus on the incidence of neuropathic pain rather than on underlying neuropathological processes [18]. Notwithstanding this orientation, the first step here will be to examine the changes in peripheral nerves that result from injury.

Understanding pathophysiology arising from peripheral nerve trauma is further complicated by the various types of trauma, which consist largely of a total cut of a peripheral nerve, a partial cut, an event-triggered compression, a slowly-developing compression (such as from a tumour), and a degeneration of nerve cells or of neural support cells and neural support tissues. Total and partial cuts as well as event-triggered compression can occur as a result of an accident, a violent act or surgery. Degeneration can be induced, for example, by chemotherapy. Physical, surgical and chemical events provide a start time for medical care [4, 12, 23]. As post-herpetic neuralgia is associated with a time-locked event, which is the start of the symptoms of shingles, this type of neuropathic pain is included in this overview [24-26], as would be any other infection-or inflammation-induced neuropathy that is time-locked.

Multiple pathophysiological, neurochemical, and anatomical changes are triggered by peripheral nerve injury, whereby a simple focal peripheral nerve injury unleashes a range of peripheral as well as central nervous system processes that contribute to persistent pain and abnormal sensation. Repair mechanisms of neural tissues in response to injury, and the reaction of adjacent tissues to injury lead to a state of hyperexcitability in primary afferent nociceptors [27], a phenomenon termed peripheral sensitization. In turn, central neurons innervated by such nociceptors undergo dramatic functional changes including a state of hyperexcitability termed central sensitization [28].

Normally these sensitization phenomena extinguish as the tissue heals and inflammation subsides. However, when primary afferent function is altered in an enduring way by injury or disease of the peripheral nerves, these processes persist, become chronic and may be highly resistant to treatment.

Much of what has been learned regarding the pathophysiology of injury causing neuropathic pain has come from animal studies. Human laboratory studies, although limited in number, support the idea that the pathophysiological mechanisms discovered in animal models are valuable and relevant to our understanding of human neuropathic pain [29, 30]. There are several animal models of peripheral neuropathic pain, largely based on the types of primary injury or trauma that lead to peripheral neuropathic pain in humans [31-39].

Effects of peripheral nerve injury in these models include major changes in the properties of nerve cells and their support cells, particularly the Schwann cells. Changes in sensory nerve cells include spontaneous ectopic action potential generation, persisting sensitization of sensory nerve cell peripheral terminals, and increased release of excitatory neurotransmitters from their central nerve terminals [27, 40-42]. Other changes in sensory nerve cells include a change in function [43], changes in the expression of cell constituents such as sodium channels [44] and changes in the expression of neurotransmitters [45, 46], including *de novo* expression of substance P in large fibre, non-nociceptive sensory neurons [40]. While there is currently no universally effective treatment for peripheral neuropathic pain, one that is included in the algorithm of treatment regimens is gabapentin [16, 21], which it thought to act by reducing the release of excitatory neurotransmitters from the central terminals of sensory neurons [47, 48]; this action is thought to occur through inhibition of the influx of calcium into the nerve terminals, which is necessary for synaptic release mechanisms [49].

Two major hypotheses have been put forward to explain the role of sensory neurons in the generation of neuropathic pain [27]. One is the classic "excitable nociceptor hypothesis", which implicates a reduced response threshold in nociceptive small C-fibre sensory neurons. Another hypothesis suggests ectopic impulse activity is generated in low threshold mechanoreceptor large fibre Aβ sensory neurons and that this activity is abnormally "amplified" in the spinal cord by central sensitization [27, 50]. The hypothesis implicating C-fibre sensory neurons in causing neuropathic pain is mainly based on the abnormal spontaneous activity observed in these neurons. However, several studies have shown that large fibre sensory neurons undergo significant changes in their electrophysiological properties as well as other phenotypic changes, such as the expression of substance P [51-53]. Further studies have shown that large fibre Aβ neurons, not the small C-fibre sensory neurons might be the major drivers of stimulus-evoked neuropathic pain and tactile allodynia [54].

Irrespective of these outcomes of peripheral nerve injury, which manifest in established neuropathic pain, the focus of the present overview is on the mechanisms and processes that lead to these outcomes. A consensus is that once these processes are set in place the resulting neuropathic pain is relatively refractory to medical treatment. The present overview therefore addresses the initial changes in peripheral nerve cells, neural support cells and neural support tissues that lead to long-term or permanent pathology, adverse health outcomes or disability.

5. Triggering events and cellular processes leading to adverse outcomes

Specific mechanisms triggered by injury to sensory nerve cells vary depending on the site of the trauma and the type of trauma [55]. There is also a reported difference in the restorative and degenerative processes activated between immature and mature rat models [56].

A cut to an axon induces an immediate influx of calcium, which disrupts the ionic balance of the nerve cell and initiates transport of a number of intracellular and extracellular chemicals to the nerve cell body in the dorsal root ganglion [57], contributing to spontaneous ectopic activity in small diameter normally nociceptive neurons and/or in the large diameter normally non-nociceptive neurons (summarized in [18, 27, 58]). Other changes are also induced in nerve cell bodies in the dorsal root ganglia when injury occurs to the axons, including chromatolysis, displacement of the cell nucleus, cell shrinkage and a decrease in axonal transport [59].

Abrupt non-cutting trauma can also cause injury to axons, which then begin to degenerate over the next several days [60, 61], leading to cell death in a delayed, progressive process [62]. Degeneration of the central and peripheral terminal projections of the damaged and dying nerve cells leads to sprouting of the terminal projections of neighbouring undamaged nerve cells, constituting a morphological reorganization of the spinal cord neuronal circuitry [63] and perhaps also of the peripheral innervations of tissues.

Trauma to the axon also alters the normal flow of proteins orthogradely toward the cell body in the dorsal root ganglion, and retrogradely toward the peripheral terminal along intracellular filaments. This alters the information arriving to the protein assembly mechanisms in the cell body and this leads to changes gene expression of proteins. Changes in gene expression follow a temporally specific pattern [64], indicating that different cellular contents are produced at varying times following trauma to a peripheral nerve. In addition to changes in the expression of proteins there are also changes in the cellular distribution of these proteins, particularly of sodium and calcium channels [65], which are critically involved in neuronal excitability and conduction. Changes in gene expression and generation of specific proteins are particularly relevant to secondary injury including changes in the expression of neurotransmitters, such as substance P, of trophic factors such as brain-derived neurotrophic factor and other factors, as well as kinases and other degradative enzymes [45, 66].

Changes in expression and distribution of calcium channels have received particular attention. Expression of the α-2-δ subunit of voltage-gated calcium channels is increased in neuropathic pain models and this increase correlates with the onset and duration of pain scores [67]. The anticonvulsant, gabapentin, and its derivative, pregabalin, are both reported to bind to the α-2-δ subunit of voltage-gated calcium channels and inhibit transport of this subunit to nerve terminals [68]. Gabapentin and pregabalin have both shown clinical efficacy for treating chronic neuropathic pain in humans [16, 69, 70].

Changes also occur in Schwann cells. Any changes in the supporting Schwann cells are important to understand because of their pivotal role in sustaining the physiological properties of peripheral nerve cells and because they are involved in a degenerative process termed 'Wallerian degeneration' [71]. Schwann cells are the glial cells of the peripheral nervous system, and include myelinating cells and non-myelinating satellite cells. Dysfunction of Schwann cells is at the basis of several peripheral nerve disorders, such as Guillain-Barré disease and Charcot-Marie-Tooth disease [72-74]. As the Schwann cells decompose so do the myelin sheaths. The products of this decomposition trigger proliferation of new undifferentiated Schwann cells that align along the Bungner's bands that constitute the tubes within which the nerve bundles are contained, along with their support cells. Toll-like receptors are strongly induced by axotomy, they are critically involved in degeneration [75.76] and they lie at the crossroads of peripheral nerve pathology and pain [77].

Changes also occur in other non-neuronal cells, affected by axonal degeneration and Wallerian degeneration of axons. Macrophages and lymphocytes as well as immune cells from the blood infiltrate dorsal root ganglion cells and are attracted to the site of nerve damage [71, 78]. This chemical milieu contains many components that impact on all cell types, including neurotrophic factors such as brain-derived neurotrophic factor, glial-derived neurotrophic factor, nerve growth factor and neurotrophin-3 [79-82], pro-and anti-inflammatory cytokines such as tumor necrosis factor (TNF-α), interleukin-1α (IL-1α), interleukin-1β (IL-1β), interleukin-6 (IL-6), interleukin-10 (IL-10) and leukemia inhibitory factor [83, 84] and the chemokine monocyte chemoattractant protein-1 [85]. Axon-promoting chemicals are thought to provide support for regrowing axons [83]. Some components are thought to cause increased excitability of undamaged axons as a cause of neuropathic pain [86]. Even cutting ventral roots has been shown to lead to neuropathic pain behaviour in rats [87], presumably due to the migration of chemicals from degenerating nerve and neural support cells associated with motor control to intact neighbouring sensory neurons [88-90].

Peripheral nerves travel alongside blood vessels and trauma to the nerve often physically disrupts the blood-nerve barrier, allowing the milieu of chemicals produced by axon and Schwann cell degeneration to enter the bloodstream, which then carries these chemicals to remote parts of the body, including direct access to uninjured dorsal root ganglia, the enteric nervous system and the central nervous system [91]. This tends to carry peripheral nerve injury to remote sites beyond the site of a primary injury.

6. Processes and mechanisms of the changes in spinal cord resulting from peripheral nerve injury

Although the focus of the present overview is on peripheral nerve injury, changes in sensory nerve cells produce changes in spinal cord nerve cells, glial cells and other neural support cells, and it is generally held that some of the outcomes of peripheral nerve injury may be brought about as a result of changes in the spinal cord. For example, peripheral nerve injury produces excessive excitation and activation of sensory neurons, which causes excessive release of glutamate, substance P and other neurotransmitters from their central terminals in the spinal cord and brain stem. Besides leading to central sensitization, released neuropeptides regulate gene expression and therefore the phenotype of neurons in the spinal cord and brain stem [92]. Outcomes of peripheral nerve injury in animal models have been reported to include major changes in the properties of spinal neurons and spinal neural support cells and neural support tissues [93-95]. Peripheral nerve injury has also been reported to produce changes in the processing of sensory information at the spinal and supraspinal levels [96-101].

As a result of acute peripheral nerve injury, discharge from both damaged and adjacent intact primary afferent fibres becomes abnormal. This modified afferent drive in turn has been reported to elevate the excitability and discharge of dorsal horn neurons [43, 102] and to induce changes in sensory processing at the level of the spinal dorsal horn including altered calcium-dependent signal transduction mechanisms [103], a shift in anion gradient [104], microglial activation [105], decreased inhibitory mechanisms [58], apoptosis [106] and others. It is thought that this modified sensory processing at the level of the spinal dorsal horn, termed central sensitization [50, 107, 108] or long term potentiation [109], contributes to neuropathic pain by exacerbating excitatory transmission to supraspinal structures.

http://dx.doi.org/10.5772/58421

In view of the evidence that central sensitization in humans is maintained by a constant barrage of synaptic activity from primary afferent nerve cells [58, 110] it is important in understanding mechanisms of neuropathic pain to understand the properties of primary sensory neurons and how they adapt to or change in response to nerve injury. It is at these properties, particularly at the processes that are involved in the initial stages of neuropathology, that early pharmaceutical intervention can be aimed.

7. Standard treatment following peripheral nerve injury

For complete cut of a peripheral nerve a standard procedure is nerve repair or nerve grafting, but this approach remains suboptimal and is usually performed long after the cut has occurred. Axons have the capacity to regrow, but this is often incomplete or the regeneration misses the original tissue target. As a result there has been a wealth of research on mechanisms of regeneration and respective treatment modalities. Yet, outcome generally remains poor.

To make matters worse, leaders in the field of pain have suggested that "there is little evidence that chronic postsurgical pain can be prevented" and they cite papers such as those by Kehlet et al. [23], Gartner et al. [111] and Katz and Seltzer [112]. In fact, this assessment might just be correct, as the concept of 'prevention' of postsurgical chronic pain is usually embedded in the concept of peri-surgical anaesthesia [113-115] rather than control of the degenerative processes that lead to secondary injury to nerve cells and Schwann cells in peripheral nerves. This latter approach is not included in steps to prevent postsurgical pain, let alone a rationalized, combination therapy based on a timed sequence of pharmaceutical interventions aimed at reducing or preventing the processes involved in secondary injury to these cells. The consensus therefore teaches away from this rationalized approach toward techniques that have been proven by evidence to be relatively ineffective, condemning multiples of thousands each year to a life of unending refractory pain.

Numerous efforts have been made to develop new and effective drugs and other approaches to treat neuropathic pain. Some treatments have been found to have beneficial effects. These include treatment with multimodal analgesics [116], anticonvulsants [16, 21], botulinum toxin [117, 118], peripheral nerve electrical stimulation [119, 120], as have invasive approaches such as spinal cord stimulation [121] and administration of stem cells [122]. However, these are aimed at treating existing pain. What are not being explored are new treatments to prevent the onset of neuropathic pain or any other of the sensory, motor of autonomic adverse sequelae of peripheral nerve injury.

8. Limitations to incentives to prevent secondary injury to peripheral nerves

Trauma to peripheral nerves is not considered life threatening, whether physical, chemical, metabolic or surgical. As a result there is limited incentive to pursue medical interventions, methods and procedures to reduce or prevent secondary peripheral nerve injury that results from trauma. For example, there is no appreciation of immediacy in medical intervention. Research on medical intervention for neuropathic pain focuses on treatment of an existing condition, once a complete diagnosis has been made. Complete or correct diagnosis can take weeks, months or even years. Standard treatment following peripheral nerve trauma typically involves drugs that reduce the pain intensity. Thus, in the pursuit of new drugs, neuropathic pain resulting from peripheral nerve trauma is becoming understood in terms of a static, or an established, condition. The processes that are involved in the initial pathophysiology of peripheral nerve injury remain poorly understood and few efforts are being made to understand or to intervene in these processes.

A second limit to incentive to understanding the pathophysiology of peripheral nerves is that a major focus of research has been on the changes in the spinal cord and other central nervous system structures that result from peripheral nerve trauma [123]. As a result, much of the research on the pathophysiology of neurons and neural support cells in neuropathic pain has focused on changes in the spinal cord, where the predominant concepts are 'central sensitization' or 'long-term potentiation' [108, 124] and 'neuroplasticity' [125], as mechanisms underlying the pain. While this is important, it has tended to shift focus away from the changes in peripheral nerves and the role of primary afferent drive in the mechanisms of neuropathic pain [27, 43, 102].

There is a rich literature pertaining to pharmacological treatments for peripheral neuropathic pain as an existing condition, summarized in a number of thorough reviews [16, 20-22, 126]. This abundant literature is largely due to the fact that neuropathic pain is a particularly debilitating type of chronic pain, yet it remains refractory to medical treatment in a large number of patients. Despite this huge medical need, there is little information or drive with regard to reducing or preventing the development of peripheral neuropathic pain, which is the focus of the present overview.

9. New approaches to minimize secondary injury following peripheral nerve injury

Despite abundant information regarding the processes and mechanisms of the changes in the periphery and in the spinal cord, there appears to be little effort being made to understand these processes and mechanisms or to develop new therapeutics to prevent trauma-induced secondary injury in peripheral nerves. Steps to reduce or to prevent the development of neuropathic pain are not generally considered in medical practice, other than steps to avoid traumatic events. Prevention, from a medical intervention standpoint, is not found in consensus statements.

Novel therapies under development retain the focus on treatment of an existing symptom. Steps to reduce or prevent secondary nerve injury following peripheral nerve trauma through immediate medical intervention do not appear in major national statements or reports, such as the US National Pain Care Policy Act of 2009, the 2011 Annual Report of the Chief Medical

http://dx.doi.org/10.5772/58421

Officer of the United Kingdom, or the 2011 report of the US Institute of Medicine, "Relieving Pain in America. A Blueprint for Transforming Prevention, Care, Education and Research".

Given the enormous impact of trauma-induced peripheral neuropathology and its sequelae on individuals, on families, on the healthcare system and on the economy, and the enormous social impact specifically of abuse of pain-relieving drugs, this presents an opportunity to exploit the limited knowledge we have regarding mechanisms underlying the secondary injury to peripheral nerves following traumatic events, to develop effective medical intervention to reduce or prevent the secondary processes that lead to peripheral nerve pathology.

Some insights into possible therapeutic approaches have come from animal studies, which have shown that peripheral neuropathy resulting from physical trauma is preventable when immediate and appropriate therapy is introduced [127] but is not preventable when treatment is delayed [128-131]. There is limited evidence that immediate or at least early medical intervention may have beneficial effects on long-term outcome. For example, immediate but not prolonged treatment with either an NK-1 receptor antagonist [130] or with progesterone [129] has been shown to have long-term benefit in an animal model of peripheral nerve injury. In fact, a recent study on early treatment with peripheral nerve stimulation of soldiers on the battlefield has reported improved functionality and opioid use reduction [120], a point well made as extremity trauma is a relatively more common medical condition in battle now because of advances in body armour [132].

As described above, due to the plethora of mechanisms that are triggered by event-related trauma to peripheral nerves combinatorial approach to reducing or preventing secondary injury to peripheral nerves may be necessary. There is presently no therapeutic approach to prevent or reduce the adverse health conditions or disability that result from trauma-induced damage to peripheral nerve cells, peripheral neural support cells or peripheral neural support tissues. There is little research directed at translation of new discoveries from animal studies to preventing or reducing peripheral neuropathology in humans and, as a result, until this invention there is little evidence or indication that this medical need will be met.

10. Recommendations for future research.

If new, appropriate therapeutics are to be developed in order to prevent or limit the disability that ensues from peripheral nerve injury, such therapeutics will have to target the biochemical and metabolic processes triggered by trauma, and therefore research is needed to understand these processes beyond what has been described in this overview. Further, from this understanding novel targets need to be identified that offer opportunities to develop novel therapeutics.

Injury triggers cascades of cellular, biochemical and metabolic processes, some of which tend to return nerve cells, neural support cells and neural support tissues toward normal function and cell health. Some changes tend to drive nerve cells, neural support cells and neural support tissues toward loss of cell function or cell death. The former group of processes is considered to be restorative or adaptive processes. The latter group of processes is considered to be degenerative or maladaptive. The eventual outcome at the cellular and tissue levels is determined by or results from the balance of all the restorative and degenerative processes triggered by or resulting from the primary injury and its sequelae. Indeed, the damage caused by these secondary processes can be as serious and extensive as, or even more serious and extensive than, that caused by the primary trauma. Secondary processes also progress over time so that injury and damage can continue over the days, weeks and even months after the initial injury. Further, the secondary processes can also progress spatially so that injury and damage can spread spatially and manifest at sites remote from the site of the primary trauma to other peripheral nerves.

This balance can be tipped toward normal function and health by appropriate pharmaceutical intervention at the appropriate time. This can be achieved because of the chemical nature or basis of the restorative and degenerative processes occurring at the cellular, biochemical and metabolic levels.

As indicated above, there are many targets or points of entry for pharmaceutical promotion, facilitation or potentiation of restorative processes to tip this balance toward function and health, and there are many targets or points of entry for pharmaceutical inhibition, lessening or blocking of degenerative processes that tip this balance away from function and health toward loss of function, adverse health conditions or disability. It is recommended, then, that future research focus on understanding these mechanisms and identifying potential targets for development of new therapeutic approaches.

11. Conclusions

Conventional or standard treatment of trauma typically consists of minimizing the symptoms of the immediate, or primary, traumatic injury. With conventional or standard methods and treatments, attempts are made to minimize these immediate symptoms. Standard treatment for any persisting loss of function or disability that results from the initial trauma is typically treated by rehabilitation, which is usually initiated after there has been overt recovery from the traumatic event itself. Initiation of rehabilitation typically comes weeks or even months later, when the adverse health conditions or disability are clear and obvious.

In significant contrast, it is suggested here that in addition to standard emergency or critical care at the time of an accident or trauma specific actions be directed toward mitigating or ameliorating the sequelae of post-trauma effects that are an indirect result of a primary trauma and that are expressed as a result of the balance of restorative and degenerative processes.

Presently, there is a gap in medical care between standard practice to treat a primary injury or damage at the time of trauma, and standard practice to rehabilitate. The suggestion here is to address this gap by understanding and promoting processes that drive toward recovery and restoration of cell health and function and at the same time inhibiting processes that drive toward loss of cell function and cell death.

http://dx.doi.org/10.5772/58421

Significantly, the point of differentiation between conventional or standard methods and the present position is the difference between the treatment of the symptoms of the primary injury, and formulations, methods and procedures taken at or about the time of trauma to prevent or lessen damage from the secondary sequelae that may or are likely to occur.

The cascades of mechanisms leading to secondary injury are triggered within minutes to hours, yet continue to occur over the ensuing days and weeks. As a result, symptoms of secondary injury manifest over such periods, and the present position argues to reduce or prevent the manifestation or expression of these symptoms of secondary injury, which are known on the basis of incidence studies to occur.

There is a general acceptance that disability results from trauma. Incidence studies indicate that a certain number of people in a population will go on to develop disability following trauma of any given type. Medical attention has not typically been directed at reducing these numbers, or preventing them altogether. Instead, it tends to be directed at saving life and addressing the immediate condition and symptoms. Yet, much of the disability that ensues as a result of trauma is brought about by secondary injury processes, largely biochemical, which can be modified by appropriate pharmaceutical intervention. Trauma-induced disability can thus be considered an unaddressed medical need. The present position here is that the number of people who go on to develop disability following trauma can be reduced. Further, the severity of disability of those that develop a disability can be reduced. The scope and the spirit of the present overview are directed toward this unaddressed need, both by reducing the number of victims of trauma that go on to develop adverse health conditions and disability, as well as by reducing the severity of disability in those who are left with trauma-induced health conditions.

Author details

James L. Henry*

Address all correspondence to: james.henry@sapnalifesciences.ca

Sapna Life Sciences Corp., ON, Canada

The author declares no conflict of interest.

References

[1] Torrance N, Smith BH, Bennett MI, Lee AJ. The epidemiology of chronic pain of predominantly neuropathic origin. Results from a general population survey. J Pain. 7:281-289, 2006.

[2] Ciaramitaro P, Mondelli M, Logullo F, Grimaldi S, Battiston B, Sard A, Scarinzi C, Migliaretti G, Faccani G, Cocito D. Traumatic peripheral nerve injuries: epidemiological findings, neuropathic pain and quality of life in 158 patients. J Periph Nerv Syst 15:120-127, 2010.

[3] Mailis A, Umana M, Feindel CM. Anterior intercostal nerve damage after coronary artery bypass graft surgery with use of internal thoracic artery graft. Ann Thorac Surg. 69: 1455-1458, 2000.

[4] Macrae WA. Chronic post-surgical pain: 10 years on. Br J Anaesthes 101: 77-86, 2008.

[5] Meyerson J, Thelin S, Gordh T, Karisten R. The incidence of chronic post-sternotomy pain after cardiac surgery-a prospective study. Acta Anaesthesiol Scand 45: 940-944, 2001.

[6] Brandsborg B, Nikolajsen L, Kehlet J, Jensen TS. Chronic pain after hysterectomy. Acta Anaesthesiol Scand 52: 327-331, 2008.

[7] Kainu JP, Sarvela J, Tiippana E, Halmesmaki E, Korttila KT. Persistent pain after caesarean section and vaginal birth: a cohort study. Int J Obstetric Anes 19: 4-9, 2010.

[8] Nikolajsen L, Sorensen JC, Jensen TS, Kehlet H. Chronic pain following caesarean section. Acta Anaesthesiol Scand 48: 111-118, 2004.

[9] Bennett MI, Rayment C, Hjermstad M, Aass N, Caraceni A, Kaasa S. Prevalence and aetiology of neuropathic pain in cancer patients: a systematic review. Pain 153: 359-365, 2012.

[10] Kanbayashi Y, Hosokawa T, Okamoto K, Konishi H, Otsuji E, Yoshikawa T, Takagi T, Taniwaki M. Statistical identification of predictors for peripheral neuropathy associated with administration of bortezomib, taxanes, oxaliplatin or vincristine using ordered logistic regression analysis. Anticancer Drugs. 21:877-881, 2010.

[11] Kuroi K, Shimozuma K, Ohashi Y, Hisamatsu K, Masuda N, Takeuchi A, Aranishi T, Morita S, Ohsumi S, Hausheer FH. Prospective assessment of chemotherapy-induced peripheral neuropathy due to weekly paclitaxel in patients with advanced or metastatic breast cancer (CSP-HOR 02 study). Support Care Cancer. 17: 1071-1080, 2009.

[12] Weimer LH, Sachdev N. Update on medication-induced peripheral neuropathy. Curr Neurol Neurosci Reports 9: 69-75, 2009.

[13] Davenport JG, Farrell DF, Sumi M. "Giant axonal neuropathy" caused by industrial chemicals: neurofilamentous axonal masses in man. Neurol 26: 919-923, 1976.

[14] Borgens RB, Liu-Snyder P. Understanding secondary injury. Quart Rev Biol 87L 89-127, 2012.

[15] Vink R, Nimmo AJ. Multifunctional drugs for head injury. Neurotherapeut 6: 28-42, 2009.

[16] Moulin DE, Clark AJ, Gilron I, Ware MA, Watson CP, Sessle BJ, Coderre T, Morley-Forster PK, Stinson J, Boulanger A, Peng P, Finley GA, Taenzer P, Squire P, Dion D, Cholkan A, Gilani A, Gordon A, Henry J, Jovey R, Lynch M, Mailis-Gagnon A, Panju A, Rollman GB, Velly A. Pharmacological management of chronic neuropathic pain – consensus statement and guidelines from the Canadian Pain Society. Pain Res Manage 12: 13-21, 2007.

[17] Kaas JH, Collins CE. Anatomic and functional reorganization of somatosensory cortex in mature primates after peripheral nerve and spinal cord injury. Adv Neurol 93: 8-95, 2003.

[18] Costigan M, Scholz, J, Woolf CJ. Neuropathic pain: a maladaptive response of the nervous system to damage. Ann Rev Neurosci 32: 1-32, 2009.

[19] Finnerup NB, Sindrup SH, Jensen TS. Chronic neuropathic pain: mechanisms, drug targets and measurement. Fund Clin Pharmacol 21: 129-136, 2007.

[20] Dworkin RH, O'Connor AB, Backonja M, Farrar JT, Finnerup NB, Jensen TS, Kalso EA, Loeser JD, Miaskowski C, Nurmikko TJ, Portenoy RK, Rice AS, Stacey BR, Treede RD, Turk DC, Wallace MS. Pharmacologic management of neuropathic pain: evidence-based recommendations. Pain 132:237–251, 2007.

[21] Gilron I, Watson PN, Cahill CM, Moulin DE. Neuropathic pain: a practical guide for the clinician. Can Med Assoc J 175: 265-275, 2006.

[22] Vorobeychik Y, Gordin V, Mao J, Chen L. Combination therapy for neuropathic pain. A review of current evidence. CNS Drugs 25: 1023-1034, 2011.

[23] Kehlet H, Jensen TS, Woolf CJ. Persistent postsurgical pain: risk factors and prevention. Lancet 367: 1618–1625, 2006.

[24] Watson CP, Oaklander AL. Postherpetic neuralgia. Pain Pract 2: 295-307, 2002.

[25] Opstelten W, McElhaney J, Weinberger B, Oaklander AL, Johnson RW. The impact of varicella zoster virus: chronic pain. J Clin Virol 48 Suppl 1: S8-S13, 2010.

[26] Lapolla W, Digiorgio C, Haitz K, Magel G, Mendoza N, Grady J, Lu W, Tyring S., Incidence of postherpetic neuralgia after combination treatment with gabapentin and valacyclovir in patients with acute herpes zoster: open-label study. Arch Dermatol 147: 901-907, 2011.

[27] Devor M. Ectopic discharge in Aβ afferents as a source of neuropathic pain. Exp. Brain Res. 196: 115-128, 2009.

[28] Woolf CJ, Salter MW. Neuronal plasticity: increasing the gain in pain. Science 288: 1765-1769, 2000.

[29] Torebjörk E. Human microneurography and intraneural microstimulation in the study of neuropathic pain. Muscle Nerve. 16: 1063-1065, 1993.

[30] Ørstavik K, Weidner C, Schmidt R, Schmelz M, Hilliges M, Jørum E, Handwerker H, Torebjörk E. Pathological C-fibres in patients with a chronic painful condition. Brain. 126: 567-578, 2003.

[31] Bennett GJ, Xie YK. A peripheral mononeuropathy in rat that produces disorders of pain sensation like those seen in man. Pain 33, 87-107, 1988.

[32] Seltzer Z, Dubner R, Shir Y. A novel behavioral model of neuropathic pain disorders produced in rats by partial sciatic nerve injury. Pain 43: 205-218, 1992.

[33] Kim SH, Chung JM. An experimental model for peripheral neuropathy produced by segmental spinal nerve ligation in the rat. Pain 50: 355-363, 1992.

[34] Mosconi T, Kruger L. Fixed-diameter polyethylene cuffs applied to the rat sciatic nerve induce a painful neuropathy: ultrastructural morphometric analysis of axonal alterations. Pain 64: 37-57, 1996.

[35] Malmberg AB, Basbaum AI. Partial sciatic nerve injury in the mouse as a model of neuropathic pain: behavioral and neuroanatomical correlates. Pain 76: 215-222, 1998.

[36] Sung B, Na HS, Kim YI, Yoon YW, Han HC, Nahm SH, Hong SK. Supraspinal involvement in the production of mechanical allodynia by spinal nerve injury in rats. Neurosci Lett 246: 117-119, 1998.

[37] Decosterd I, Woolf CJ. Spared nerve injury: an animal model of persistent peripheral neuropathic pain. Pain 87: 149-158, 2000.

[38] Lee BH, Won R, Baik EJ, Lee SH, Moon CH. An animal model of neuropathic pain employing injury to the sciatic nerve branches. Neuroreport 11: 657-661, 2000.

[39] Vadakkan KI, Jia YH, Zhuo M. A behavioral model of neuropathic pain induced by ligation of the common peroneal nerve in mice. J Pain 6: 747-756, 2005.

[40] Malcangio M, Ramer MS, Jones MG, McMahon SB. Abnormal substance P release from the spinal cord following injury to primary sensory neurons. Eur. J. Neurosci. 12: 397-399, 2000.

[41] Gardell LR, Vanderah TW, Gardell SE, Wang R, Ossipov MH, Lai J, Porreca F. Enhanced evoked excitatory transmitter release in experimental neuropathy requires descending facilitation. J. Neurosci. 23: 8370-8379, 2003.

[42] Katz EJ, Gold MS. Inflammatory hyperalgesia: a role for the C-fiber sensory neuron cell body? J Pain. 7: 170-178, 2006.

[43] Pitcher GM, Henry JL. Nociceptive response to innocuous mechanical stimulation is mediated via myelinated afferents and NK-1 receptor activation in a rat model of neuropathic pain. Exp. Neurol. 186: 173-197, 2004.

[44] Amaya F, Wang H, Costigan M, Allchorne AJ, Hatcher JP, Egerton J, Stean T, Morisset V, Grose D, Gunthorpe MJ, Chessell IP, Tate S, Green PJ, Woolf CJ. The voltage-

gated sodium channel Na(v)1.9 is an effector of peripheral inflammatory pain hypersensitivity. J. Neurosci. 26: 12852-12860, 2006.

[45] Noguchi K, Kawai Y, Fukuoka T, Senba E, Miki K. Substance P induced by peripheral nerve injury in primary afferent sensory neurons and its effect on dorsal column nucleus neurons. J Neursoci 15: 7633-7643, 1995.

[46] Hofmann HA, De VJ, Siegling A, Spreyer P, Denzer D. Pharmacological sensitivity and gene expression analysis of the tibial nerve injury model of neuropathic pain. Eur. J. Pharmacol. 470: 17-25, 2003.

[47] Yang RH, Xing JL, Duan JH, Hu SJ. Effects of gabapentin on spontaneous discharges and subthreshold membrane potential oscillation of type A neurons in injured DRG. Pain 116: 187-193, 2005.

[48] Dooley DJ, Taylor CP, Donevan S, Feltner D. Ca2+channel alpha2delta ligands: novel modulators of neurotransmission. Trends Pharmacol. Sci. 28: 75-82, 2007.

[49] Fink K, Meder W, Dooley DJ, Gothert M. Inhibition of neuronal Ca[2+) influx by gabapentin and subsequent reduction of neurotransmitter release from rat neocortical slices. Br. J. Pharmacol. 130: 900-906, 2000.

[50] Campbell JN, Raja SN, Meyer RA, Mackinnon SE. Myelinated afferents signal the hyperalgesia associated with nerve injury. Pain 32: 89-94, 1988.

[51] Han HC, Lee DH, Chung JM. Characteristics of ectopic discharges in a rat neuropathic pain model. Pain 84: 253-261, 2000.

[52] Liu CN, Wall PD, Ben-Dor E, Michaelis M, Amir R, Devor M. Tactile allodynia in the absence of C-fiber activation: altered firing properties of DRG neurons following spinal nerve injury. Pain 85: 503-521, 2000.

[53] Tal M, Wall PD, Devor M. Myelinated afferent fiber types that become spontaneously active and mechanosensitive following nerve transection in the rat. Brain Res. 824: 218-223, 1999

[54] Zhu YF, Wu Q, Henry JL. Changes in functional properties of A-type but not C-type sensory neurons in vivo in a rat model of peripheral neuropathy. J Pain Res 5:175-192, 2012.

[55] Wong J, Oblinger MM. A comparison of peripheral and central axotomy effects on neurofilament and tubulin gene expression in rat dorsal root ganglion neurons. J Neurosci 10: 2215-2222, 1990

[56] Whiteside G, Doyle CA, Hunt SP, Munglani R. Differential time course of neuronal and glial apoptosis in neonatal rat dorsal root ganglia after sciatic nerve axotomy. Wur J Neurosci 10: 3400-3408, 1998.

[57] George FB, Glass JD, Griffin JW. Axotomy-induced axonal degeneration is mediated by calcium influx through ion-specific channels. J Neurosci 15: 6445-6452, 1995.

[58] Campbell JN, Meyer RA. Mechanisms of neuropathic pain. Neuron 52: 77-92, 2006.

[59] Blumberg H, Janig W. Changes in unmyelinated fibers including sympathetic post-ganglionic fibers of a skin nerve after peripheral neuroma formation. J Auton Nerv Syst 6: 173-183, 1982.

[60] Gillatt RW, Jhorth RJ. Nerve conduction during Wallerian degeneration in the baboon. J Neurol Neuruosurg Psychiat 35: 335-341, 1972.

[61] Chaudhry V, Cornblath DR. Wallerian degeneration in human nerves: serial electrophysiological studies. Muscle Nerve 15: 687-693, 1992.

[62] Tandrup T, Woolf CJ, Coggeshall RE. Delayed loss of small dorsal root ganglion cells after transection of the rat sciatic nerve. J Comp Neurol 26: 172-180, 2006.

[63] Woolf CJ, Shortland P, Coggeshall RE. Peripheral nerve injury triggers central sprouting of myelinated afferents. Nature 355: 75-78, 1992.

[64] Kim D, Figueroa KW, Li K-W, Boroujerdi A, Yolo, T, Luo ZD. Profiling of dynamically changed gene expression in dorsal root ganglia post peripheral nerve injury and a critical role of injury-induced glial fibrillary acetic protein in maintenance of pain behaviors. Pain 143: 114-122, 2009.

[65] Gold MS, Weinreich D, Kim C-S, Wang R, Treanor J, Porreca F, Lai J. Redistribution of NaV1.8 in uninjured axons enables neuropathic pain. J Neurosci 23: 158-166, 2003.

[66] Fukuoka T, Kondo E, Dai Y, Hashimoto N, Noguchi K. Brain-derived neurotrophic factor increases in the uninjured dorsal root ganglion neurons in selective spinal nerve ligation model. J Neurosci 21: 4891-5900, 2001.

[67] Luo ZD, Chaplan SR, Higuera ES, Sorkin LS, Stauderman KA, Williams ME, Yaksh TL. Upregulation of dorsal root ganglion $\alpha_2\delta$ calcium channel subunit and its correlation with allodynia in spinal nerve-injured rats. J Neurosci 15: 1868-1875, 2001.

[68] Bauer CS, Rahman W, Tran-van-Minh A, Lujan R, Dickenson AH, Dolphin AC. The anti-allodynic $\alpha_2\delta$ ligand pregabalin inhibits the trafficking of the calcium channel $\alpha_2\delta$-1subunit to presynaptic terminals in vivo. Biochem Soc Trans. 2010 Apr; 38[2]: 525-528, 2010.

[69] Gajraj NM. Pregabalin: its pharmacology and use in pain management. Anesth Analg. 105:1805-1815, 2007.

[70] Taylor CP. Mechanisms of analgesia by gabapentin and pregabalin-calcium channel alpha2-delta (Cavalpha2-delta) ligands Pain 142: 13-16, 2009.

[71] Stoll G, Jander S. Myers RR. Degeneration and regeneration of the peripheral nervous system: from Augustus Waller's observations to neuroinflammation. J Periph Nerv Syst 7: 13-27, 2002.

[72] Kleopa KA, Orthmann-Murphy J, Sargiannidou I. Gap junction disorders of myelinating cells. Rev Neurosci. 21: 397-419, 2010.

[73] Nualart-Marti A, Solsona C, Fields RD. Gap junction communication in myelinating glia. Biochim Biophys Acta. 1828: 69-78, 2013.

[74] Talukder RK, Sutradhar SR, Rahman KM, Uddin MJ, Akhter H. Guillain Barre syndrome. Mymensing Med J. 20:748-56, 2011.

[75] Boivin A, Pineau I, Barrette B, Filali M, Vallières N, Rivest S, Lacroix S. Toll-like receptor signaling is critical for Wallerian degeneration and functional recovery after peripheral nerve injury. J Neurosci 27: 12565-12576, 2007.

[76] Goethais S, Ydens E, Timmerman V, Janssens S. Toll-like receptor expression in the peripheral nerve. Glia 58: 1701-1709, 2010.

[77] Wu Q, Henry JL. Toll-like Receptors: at the intersection of osteoarthritis pathology and pain. In: Principles of Osteoarthritis 1. Ed Rothschild BM. InTech. (ISBN 979-953-307-082-6] pp. 429-444, 2012.

[78] Hu P, McLachlan EM. Macrophage and lymphocyte invasion of dorsal root ganglia after peripheral nerve lesions in the rat. Neuroscience 112: 23-38, 2002.

[79] Obata K, Noguchi K. BDNF in sensory neurons and chronic pain. Neurosci Res 55: 1-10, 2006.

[80] Merighi A, Bardoni R, Salio C, Lossi L, Ferrini F, Prandini M, Zonta M, Gustincich S, Carmignoto G. Presynaptic functional trkB receptors mediate the release of excitatory neurotransmitters from primary afferent terminals in lamina II (substantia gelatinosa) of postnatal rat spinal cord. Dev Neurobiol. 68: 457-475, 2008.

[81] Wilson-Gerwing TD, Stucky CL, McComb GW, Verge VMK. Neurotrophin-3 significantly reduces sodium channel expression linked to neuropathic pain states. Exp Neurol 213: 303-314, 2008.

[82] Quintao NL, Santos AR, Campos MM, Calixto JB. The role of neurotrophic factors in genesis and maintenance of mechanical hypernociception after brachial plexus avulsion in mice. Pain 136: 125-133, 2008.

[83] Shamash S, Reighert F, Rotshenker S. The cytokine network of Wallerian degeneration: tumor necrosis factor-alpha, interleukin-1 alpha, and interleukin-1 beta. J Neurosci 22: 3052-3060, 2002.

[84] Schafers MM, Gets C, Svensson CI, Luo ZD, Sommer C. Selective increase of tumour necrosis factor-alpha in injured and spared myelinated primary afferents after chronic constrictive injury of rat sciatic nerve. Eur J Neurosci 17: 791-804, 2003.

[85] Jeon SM, Lee KM, Cho JH. Expression of monocyte chemoattractant protein-1 in rat dorsal root ganglia and spinal cord in experimental models of neuropathic pain. Brain Res 1251: 103-111, 2009.

[86] Marchand F, Perretti M, McMahon SB. Role of the immune system in chronic pain. Nature Rev Neurosci 6: 521-532, 2005.

[87] Sheth RN, Dorsi MJ, Li Y, Murinson BB, Belzberg AJ, Griffin JW, Meyer RA. Mechanical hyperalgesia after an L5 ventral rhizotomy or an L5 ganglionectomy in the rat. Pain 96: 63-72, 2002.

[88] Zhang JM, Li H, Liu B, Brull SJ. Acute topical application of tumor necrosis factor alpha evokes protein kinase A-dependent responses in rat sensory neurons. J Neurophysiol 88: 1387-1392, 2002.

[89] Oh SB, Cho C, Miller RJ. Electrophysiological analysis of neuronal chemokine receptors. Methods 29: 335-344, 2003.

[90] Sun JH, Yang B, Donnelly DF, Ma C, LaMotte RH. MCP-1 enhances excitability of nociceptive neurons in chronically compressed dorsal root ganglia. J Neurophysiol 96: 2189-2199, 2006.

[91] Arvidson B. Distribution of intravenously injected protein tracers in peripheral ganglia of adult mice. Exp Neurol 63: 388-410, 1979.

[92] Seybold VS, Colcou LG, Groth RD, Mermelstein PG. Substance P initiates NFAT-dependent gene expression in spinal neurons. J Neurochem 97: 397-407, 2006.

[93] Moore KA, Kohno T, Karchewski LA, Scholz J, Baba H, Woolf CJ. Partial peripheral nerve injury promotes a selective loss of GABAergic inhibition in the superficial dorsal horn of the spinal cord. J. Neurosci. 22: 6724-6731, 2002.

[94] Coull JA, Boudreau D, Bachand K, Prescott SA, Nault F, Sik A, De Koninck P, De Koninck Y. Trans-synaptic shift in anion gradient in spinal lamina I neurons as a mechanism of neuropathic pain. Nature 424: 938-942, 2003.

[95] Sandkuhler J, Liu X. Induction of long-term potentiation at spinal synapses by noxious stimulation or nerve injury. Eur. J. Neurosci. 10: 2476-2480, 1998.

[96] Carlson JD, Maire JJ, Martenson ME, Heinricher MM. Sensitization of pain-modulating neurons in the rostral ventromedial medulla after peripheral nerve injury. J. Neurosci. 27: 13222-13231, 2007.

[97] Gardell LR, Vanderah TW, Gardell SE, Wang R, Ossipov MH, Lai J, Porreca F. Enhanced evoked excitatory transmitter release in experimental neuropathy requires descending facilitation. J. Neurosci. 23: 8370-8379, 2003.

[98] Porreca F, Burgess SE, Gardell LR, Vanderah TW, Malan TP, Ossipov MH, Lappi DA, Lai J. Inhibition of neuropathic pain by selective ablation of brainstem medullary cells expressing the mu-opioid receptor. J. Neurosci. 21: 5281-5288, 2001.

[99] Saade NE, Al AH, Abdel BS, Safieh-Garabedian B, Atweh SF, Jabbur SJ. Transient attenuation of neuropathic manifestations in rats following lesion or reversible block of the lateral thalamic somatosensory nuclei. Exp. Neurol. 197: 157-166, 2006a.

[100] Saade NE, Al AH, Chalouhi S, Baki SA, Jabbur SJ, Atweh SF. Spinal pathways involved in supraspinal modulation of neuropathic manifestations in rats. Pain 126: 280-293, 2006b.

[101] Suzuki R, Rahman W, Hunt SP, Dickenson AH. Descending facilitatory control of mechanically evoked responses is enhanced in deep dorsal horn neurones following peripheral nerve injury. Brain Res. 1019: 68-76, 2004.

[102] Pitcher GM, Henry JL. Cellular mechanisms of hyperalgesia and spontaneous pain in a spinalized rat model of peripheral neuropathy: changes in myelinated afferent inputs implicated. Eur. J. Neurosci. 12: 2006-2020, 2000.

[103] Wei F, Vadakkan KI, Toyoda H, Wu LJ, Zhao MG, Xu H, Shum FW, Jia YH, Zhuo M. Calcium calmodulin-stimulated adenylyl cyclases contribute to activation of extracellular signal-regulated kinase in spinal dorsal horn neurons in adult rats and mice. J Neurosci 26: 851-861, 2006.

[104] Prescott SA, Sejnowski TJ, De Koninck Y. Reduction of anion reversal potential subverts the inhibitory control of firing rate in spinal lamina I neurons: towards a biophysical basis for neuropathic pain. Mol Pain. 2:32, 2006.

[105] Coull JA, Beggs S, Boudreau D, Boivin D, Tsuda M, Inoue K, Gravel C, Salter MW, De Koninck Y. BDNF from microglia causes the shift in neuronal anion gradient underlying neuropathic pain. Nature. 438: 1017-1021, 2005.

[106] Scholz J, Broom DC, Youn DH, Mills CD, Kohno T, Suter MR, Moore KA, Decosterd I, Coggeshall RE, Woolf CJ. Blocking caspase activity prevents transsynaptic neuronal apoptosis and the loss of inhibition in lamina II of the dorsal horn after peripheral nerve injury. J Neurosci. 25: 7317-7323, 2005.

[107] Sotgiu ML, Biella G. Contribution of central sensitization to the pain-related abnormal activity in neuropathic rats. Somatosens Mot Res. 17: 32-38, 2000.

[108] Woolf CJ. Central sensitization: implications for the diagnosis and treatment of pain. Pain 152: S2-S15, 2011.

[109] Ikeda H, Kiritoshi T, Murase K. Synaptic plasticity in the spinal dorsal horn. Neurosci Res 64: 133-136, 2009.

[110] Lang PM, Schober GM, Rolke R, Wagner S, Hilge R, Offenbächer M, Treede RD, Hoffmann U, Irnich D. Sensory neuropathy and signs of central sensitization in patients with peripheral arterial disease. Pain 124: 190-200, 2006.

[111] Gartner R, Jensen MB, Nielsen J, Ewertz M, Kroman N, Kehlet H. Prevalence of and factors associated with persistent pain following breast cancer surgery. JAMA 302:1985–1992, 2009.

[112] Katz J, Seltzer Z. Transition from acute to chronic postsurgical pain: risk factors and protective factors. Expert Rev Neurotherap 9:723–744, 2009.

[113] Katz J, Clarke H. Preventive analgesia and beyond: Current status, evidence, and future directions, Clinical Pain Management: Acute Pain, 2nd edition. Edited by Macintyre PE, Rowbotham DJ, Howard R. London, Hodder Arnold Ltd, pp 154–98, 2008.

[114] Brennan TJ, Kehlet H. Preventive analgesia to reduce wound hyperalgesia and persistent postsurgical pain. Anesthesiology 103:681–683, 2005.

[115] Raja SN, Dougherty PM. Reversing tissue injury-induced plastic changes in the spinal cord: the search for the magic bullet. Reg Anesth Pain Med 25:441-444, 2000.

[116] Reuben SS, Buvanendran A. Preventing the development of chronic pain after orthopaedic surgery with preventive multimodal analgesic techniques. J Bone Joint Surg Am 89: 1343-1348, 2007.

[117] Ranoux D, Attal N, Morain F, Bouhassira D. Botulinum toxin type a induces direct analgesic effects in chronic neuropathic pain Ann Neurol 64: 274-283, 2008.

[118] Jabbari B, Machado D. Treatment of refractory pain with botulinum toxins-an evidence-based review. Pain Med 12: 1594-1606, 2011,

[119] Mobbs RJ, Nair S, Blum P. Peripheral nerve stimulation for the treatment of chronic pain. J Clin Neurosci 14:216–221, 2007.

[120] Kent M, Upp J, Spevak C, Shannon C, Buckenmaier C. Ultrasound-guided peripheral nerve stimulator placement in two soldiers with acute battlefield neuropathic pain. Anesth Analg. 2012.

[121] Miyazaki Y, Koike H, Akane A, Shibata Y, Nishiwaki K, Sobue G. Spinal cord stimulation markedly ameliorated refractory neuropathic pain in transthyretin Val30Met familial amyloid polyneuropathy. Amyloid. 18:87-90, 2011.

[122] Franchi S, Valsecchi AE, Borsani E, Procacci P, Ferrari D, Zalfa C, Sartori P, Rodella LF, Vescovi A, Maione S, Rossi F, Sacerdote P, Colleoni M, Panerai AE. Intravenous neural stem cells abolish nociceptive hypersensitivity and trigger nerve regeneration in experimental neuropathy. Pain 153: 850-861, 2012.

[123] Kuner R. Central mechanisms of pathological pain. Nature Med 16: 1258-1266, 2010.

[124] Ruscheweyh R, Wilder-Smith O, Dridla R, Liu XG, Sandkühler J. Long-term potentiation in spinal nociceptive pathways as a novel target for pain therapy. Mol Pain 7:20, 2011.

[125] Zhuo M, Wu G, Wu LJ. Neuronal and microglial mechanisms of neuropathic pain. Mol Pain 4:32, 2011.

[126] Finnerup NB, Otto M, McQuay JH, Jensen TS, Sindrup SH. Algorithm for neuropathic pain treatment: an evidence based proposal. Pain 118: 289-305, 2005.

[127] Ji RR, Kohno T, Moore KA, Woolf CJ. Central sensitization and LTP: do pain and memory share similar mechanisms? Trends Neurosci 26: 696-705, 2003.

[128] Chao PK, Lu KT, Lee YL, Chen JC, Wang HL, Yang YL, Cheng MY, Liao MF, Ro LS. Early systemic granulocyte-colony stimulating factor treatment attenuates neuropathic pain after peripheral nerve injury. PLoS One. 7[8]:e43680, 2012.

[129] Dableh LJ, Henry JL. Progesterone prevents development of a rat model of neuropathic pain: timing and duration of treatment are critical. J Pain Research 4: 91-101, 2011.

[130] Dableh LJ, Yashpal K, Henry JL. Neuropathic pain as a process: reversal of chronification in an animal model. J Pain Research 4: 315-323, 2011.

[131] Jeon Y, Kim CE, Jung D, Kwak K, Park S, Lim D, Kim S, Baek W. Curcumin could prevent the development of chronic neuropathic pain in rats with peripheral nerve injury. Curr Ther Res Clin Exp. 74: 1-4, 2013.

[132] Mercer SJ, Chavan S, Tong JL, Connor DJ, de Mello WF. The early detection and management of neuropathic pain following combat injury. J R Army Med Corps 155:94–98, 2009.

Chapter 2

Neural Differentiation of Stem Cells in Biodegradable Three-Dimensional Scaffolds – A Novel Strategy for Nerve Regeneration

Haigang Gu, Zhilian Yue and Qi Zhang

Additional information is available at the end of the chapter

http://dx.doi.org/10.5772/58559

1. Introduction

The nervous system consists of the peripheral nervous system (PNS) and the central nervous system (CNS). Most functions of the nervous system are performed by neurons, such as movement and sense. But neurons lose the proliferation ability after maturation. Under pathological conditions, injured neurons will degenerate and die. Eventually, patients will lose some of the normal functions [1]. Although astrocytes are required for neurogenesis, synaptic maturation and neuronal activity maintenance, in current opinions, neurological diseases are caused by neurodegeneration or neuronal cell death. How to stimulate the neuroregeneration is still a key challenge in both fundamental and clinical research. Thus far, scientists have made a lot of efforts to develop drugs and devices to stimulate the functional recovery after nerve injury. There are no efficient methods available to stop or reverse neurodegeneration or neuronal cell death [2].

Current strategy for peripheral nerve injuries when the gap is less than 5 mm is to join the distal and proximal stumps of the damaged nerves by microsurgery. When the gap is longer than 5 mm, direct microsurgery will cause the tension of nerve fibers. A nerve graft needs to be used to fill the gap and make the connections between the distal and proximal stumps of the damaged nerves, in order to facilitate the regenerated nerve fibers to find their targets easily during the recovery [3, 4]. Autogenous nerve grafts require a second surgery to isolate the donor nerve tissue, which often leads to second deformities and the morbidities of donor tissues. On the other hand, the quantity for nerve autografting is quite limited. Although allografts and xenografts could serve as a possible alternatives to autografts, systemic immuno-rejection remains a major concern. Immunosuppression drugs were used to inhibit systemic immuno-rejection, which will cause adverse side effects [5, 6]. Recent advances in stem cell biology and biomaterials make it possible to develop biodegradable nerve grafts for neural tissue engineering, nerve repair and regeneration. Here, we focus on stem cell-derived neural cells and biodegradable 3D neural scaffolds or conduits for nerve regeneration.

2. Support cells derived from stem cells for nerve repair

For peripheral nerve repair, axon outgrowth needs aligned Schwann cells (called astrocytes in central nervous system) to guide the orientation and to give the support. On the other hand, damaged neurons need to be replaced to perform appropriate functions. Recent progress in stem cell biology and techniques allows us to generate large quantities of functional neurons and transplantable astrocytes from stem cells [7, 8]. Stem cell-derived neural cells have been used for the studies of axon regeneration and for the treatment of neurological diseases, such as spinal cord injury (SCI), Parkinson's disease (PD), Alzheimer's disease (AD) and amyotrophic lateral sclerosis (ALS), and showed promising functional recovery in the animal models of these neurological diseases [9-12]. Transplanted stem cell-derived neurons could integrate and form functional synaptic connections with host neurons [13]. Furthermore, transplanted stem cell-derived immature astroglial cells could become mature astrocytes by forming connections with blood vessels and transplanted induced oligodendrocyte progenitor cells (iOPCs) could form myelin sheath [14, 15].

2.1. Schwann cells

In the PNS, the majority of glial cells are Schwann cells including myelinating Schwann cells and non-myelinating Schwann cells, which play essential roles for supporting normal neuronal functions and the survival and axonal regeneration of neurons after nerve injury. The myelinating Schwann cells forming myelin sheaths around axons insulate individual axon. Similar functions are performed by in the CNS oligodendrocytes. The non-myelinating cells in the PNS show similar functions with astrocytes in the CNS, which mediate the development, mechanical and metabolic support functions and promoting neuronal survival after injury. Schwann cells are involved in maintaining normal functions of PNS, including secretion and nerve extracellular matrix, nerve development and maturation, and modulation of neuromuscular junction transmission [16]. Schwann cells also produce and secret different neurotrophic factors, such as nerve growth factor (NGF), brain derived neurotrophic factor (BDNF), neurotrophic factor-3 (NT-3), glial derived neurotrophic factor (GDNF) and insulin-like growth factor (IGF) [17, 18]. All these neurotrophic factors are indispensable for neural development and regeneration. Furthermore, Schwann cells present antigens to T-lymphocytes and are involved in the clearing of myelin debris by phagocytosis.

In response to nerve injury, Schwann cells undergo proliferation and their basal lamina forming nerve conduit to support and guide axon regeneration and outgrowth. The DNA and RNA biosynthesis and up-regulation of Schwann cells could be observed as early as 2 h after injury. In 1980, Salzer and Bunge have found that direct mechanical injury is mitogenic for Schwann cells during Wallerian degeneration and Schwann cells indeed proliferation in situ after excision [19]. However, the proliferative ability of Schwann cells is low in vivo. In most cases, endogenous Schwann cells proliferation is not enough to support and guide axon regeneration and outgrowth. Extra Schwann cells need to be transplanted after nerve damage. Morrissey et al. developed a culture method that could yield up to 98% pure Schwann cells from adult rat sciatic nerve [20]. Imaizumi et al. transplanted Schwann cells to the rat model of SCI and characterized the functional recovery by electrophysiological recording. They found that Schwann cells transplantation could form new pathway across the transaction site and provide a functional recovery of SCI [21]. Guenard et al. isoloated and cultured Schwann cells from rat sciatic nerve and seeded cultured Schwann cells into semipermeable guidance channels. And then, they implanted Schwann cells-loaded nerve guidance channels into 8 mm rat sciatic nerve gap. Interestingly, they found that there was a positive correlation between the number of transplanted cells and the number of myelinated axons. Furthermore, they found that implanted Schwann cells loaded nerve guidance channels could improve the neural regenerative process [22]. Berrocal et al. implanted absorbable collagen conduits in combination with autologous SCs to a critical size defect (13 mm) in the sciatic nerve of male Fischer rats. Their results showed that absorbable collagen conduits loaded with Schwann cells significantly enhanced the regeneration of myelinated axons. The generated axons could grow into the nerve stump into the proximal and middle of the tube 4 weeks after implantation. The regeneration of myelinated axons occupied the entire length of the nerve guide 16 weeks after implantation. Functional recovery was observed in the animals who received implant treatment [23].

However, transplantation of Schwann cells is impractical for clinical application. Firstly, nerve tissue that could be used to isolate Schwann cells is limited in the patients. Secondly, Schwann cells need time to grow adequate amounts of cells for transplantation. It will take a couple of weeks, even longer. Thirdly, delayed Schwann cells transplantation often reduces the functional recovery after nerve injury compared with whose received acute application. Scientists began to seek other alternative sources for the treatment after nerve injury. Schwann cells can be generated from adult stem cells, such as mesenchymal stem cells (MSCs) and neural stem/progenitor cells (NSPCs), and pluripotent stem cells including embryonic stem cells (ESCs) and induced pluripotent stem cells (iPSCs) in vitro and in vivo. Stem cells are good alternative source for Schwann cells. Furthermore, stem cells could locally differentiate into glial cells as well as neurons after transplantation.

2.2. Mesenchymal stem cells

Mesenchymal stem cells (MSCs) can almost be found in any adult organ and can be easily harvested from patients. MSCs are capable of self-replication to many passages and can be expanded to enough cell numbers for tissue and organ regeneration. Although MSCs have been firstly harvested from the bone marrow, they actually have different properties from bone marrow stromal cells (BMSCs). BMSCs are a highly heterogeneous cell population, which includes multiple cell types with different potentials for proliferation and differentiation. On the contrary, bone marrow MSCs are a more homogenous subtype of mononuclear progenitor cells that have stem cell properties, such as self-renewal capacity and multipotency [24, 25]. Bone marrow MSCs undergo to differentiate into adipocytes, chondrocytes and osteocytes in culture. In addition, bone marrow MSCs express specific cell surface markers, such as positive for CD105, CD166, CD29 and CD44 and negative for CD14, CD34 and CD45. MSCs can also be derived from other non-marrow tissues, such as the liver and adipose, lung, peripheral blood, as well as amniotic fluid, umbilical cord blood and Wharton's jelly of the umbilical cord [26, 27]. MSCs are not only able to differentiate into mesodermal cell phenotypes but also into ectodermal lineage, Schwann cells, astroglial cells, oligodendrocytes and neurons, such as dopaminergic and purkinje neurons and have been used to treat cardiac and neurological disorders [28, 29]. Adipose derived stem cells (ADSCs) are a subtype of MSCs, which isolated from adipose tissue. Like bone marrow MSC, ADSCs are also self-renewal capacity and ability to differentiate into multiple lineages. Compared bone marrow MSC, people found that ADSCs are easier to harvest and culture for longer periods and grow faster [30]. MSCs could be induced to differentiate into neurons and Schwann cells in the regular culture vessels. Furthermore, MSCs have been induced to differentiate into neuronal-like cells expressing neuronal biomarkers, such as Tuj1 and neurofilament, and neural progenitor cells forming neurosphere-like structure in the three dimensional (3D) biodegradable scaffolds [24, 31-33]. All-trans retinoic acid is a most common drug used to initiate Schwann cell differentiation of MSCs. Forskolin, FGF2, Platelet-Derived Growth Factor (PDGF) and Neuregulin NRG1-1 are often used to force the final Schwann cell differentiation [34]. Interestingly, human MSCs can also be induced differentiated into Schwann cells. Human MSCs-derived Schwann cells have Schwann cell morphology and expression Schwann cell-specific proteins, such as p75 neurotrophin factor. Furthermore, Human MSCs-derived Schwann cells secrete several growth factors, such as hepatocyte growth factor (HGF) and vascular endothelial growth factor (VEGF) in vitro and in vivo. Transplantation of human MSCs-derived Schwann cells dramatically enhanced axonal outgrowth in an animal model of spinal cord injury [35].

Although a couple of reports showed that that MSCs can be used to generate neuronal cells, this phenomenon was recently called into question [36, 37]. Firstly, there is no evidence that neural tissues directly generated from MSCs. Secondly, the functional properties of MSCs-derived neurons have not been extensively studied, such as patch clamp recording for neuronal activities and high-performance liquid chromatography (HPLC) for neurotransmitter release. Thirdly, similar culture conditions used to induce MSCs to differentiate into neurons could also induced fibroblasts to neuronal-like cells. However, there is no doubt about clinical improvements demonstrated in animal models and patients after treatment with MSCs. People believe that these clinical improvements are from growth factors and cytokines released from MSCs. MSCs-derived growth factors and cytokines could promote neurogenesis and angiogenesis of damaged brain tissue and inhibit the process of apoptosis. Transplantation of MSCs has shown a significant functional recovery of the animal model of stroke. Nevertheless, MSCs are a good source for cell therapy.

2.3. Neural stem/progenitor cells

Previous studies showed that adult neural stem/progenitor cells (NSPCs) not only locate in neurogenic regions, the subventricular zone (SVZ) and the subgranular zone (SGZ) of the hippocampal dentate gyrus, but also locate in some non-neurogenic regions, such as cerebral cortex, cerebellum and spinal cord [7]. Multipotent CNS stem-like cells were first cultured from the adult striatum by neurosphere assay. The neurosphere culture system has been widely used to isolate and expand NSPCs under serum-free media conditions as well as with the presence of epidermal growth factor (EGF) and fibroblast growth factor-2 (FGF2). NSPCs are self-renewing, multipotent progenitors in the nervous system that could be induced to differentiate into the three phenotypes in the nervous system under appropriate condition, such as neurons, astrocytes and oligodendrocytes. Functional properties of NSPCs-derived neurons have been characterized by immunostaining and clamp patch recording. NSPCs-derived neurons have all the properties of native neurons. NSPCs have been widely used for the studies of neural development and regeneration. After transplantation, NSPCs-derived neurons could replace the dead neurons in the animal's model of neurological disorders, such as AD, PD and SCI and have shown the functional recovery in these animals' models [38, 39]. In our previous studies, we isolated and cultured NSPCs from mouse and rat brains under DMEM/F-12 medium supplement with B27 (2%) or N2 (1%), EGF (10 ng/ml) and FGF2 (10 ng/ml) by neurospheres assay. After 2-3 passages, most of neurospheres are positive to Nestin, a neural stem cell marker. We found that BrdU could be detected 16-18 h after NSPCs were cultured in the presence of BrdU. 5-bromo-2'-deoxyuridine (BrdU) is a synthetic nucleoside that is an analog of thymidine and is widely used in the detection of proliferating cells in vitro and in vivo. When NSPCs were cultured in the serum containing medium, they are easily induced to differentiate into neurons and astrocytes. These results demonstrate that NSPCs from mouse and rat brain have a high proliferative ability and multipotency. After transplantation to AD rats with fimbria-fornix transection, NSPCs could migrate into adjacent brain tissue and locally differentiate into neurons and astrocytes. Y-maze testing showed that transplanted NSPCs could improve the learning and memory in the rat model of AD [13, 40-45]. NSPCs have been cultured in a 3D bioactive scaffold derived from porcine urinary bladder matrix (UBM) for the treatment of traumatic brain injury (TBI). UBM was able to support extended proliferation and differentiation of NSPCs. After transplantation into rat TBI model, the transplants could reduce neuronal loss and white matter injury, and also significantly ameliorate motor, cognitive and memory impairments [46]. Furthermore, NSPCs-loaded PLGA scaffolds have been used for the treatment of animal model of SCI. NSPCs could differentiate into neurons and glial cells in the PLGA scaffolds after transplantation and make functional synaptic connections with proximal and distal nerve stumps. Retrograde tracking studies showed that the tracer could pass through the nerve gap and be found in the brain [47].

In the developing CNS, the initial symmetric cell division of NSPCs occurs to produce more identical stem cells and form neural tube. As development progresses, symmetric cell division is gradually replaced by asymmetric cell division which produce one stem cell and one neural precursor cell. Previous studies have identified neuronal restricted precursors (NRPs) and glial restricted precursors (GRPs) in the brain and spinal cord. These cells are more limited in their differentiation potential than NSPCs. NRPs have the tendency to differentiate into neurons and GRPs have the tendency to differentiate into glial cells. After transplantation in the model of SCI, transplanted GPCs demonstrated to differentiate into oligodendrocytes and form myelin around axons and increase the locomotor recovery [48]. Furthermore, NSPCs have been used along with Schwann cells to improve axonal regeneration. Olson et al. transplanted NSPCs and Schwann cell-loaded PLGA polymer scaffold into transected spinal cord. They found that NSPCs could differentiate into neuronal cells in the scaffold channels and NSPCs and Schwann cell-loaded PLGA polymer scaffold could facilitate axonal regeneration across the transected spinal cord [49].

The major drawback of NSPCs is that NSPCs locate in the deep of brain and spinal cord. It is almost impossible to harvest autologous NSPCs for nerve repair in clinic. Although NSPCs could be obtained from aborted embryos, ethical issue plagues their clinical application.

2.4. Induced neural cells

In 2012, John B. Gurdon and Shinya Yamanaka shared the Nobel Prize in Physiology or Medicine for their discovery that mature cells can be reprogrammed to become pluripotent. In 1962, Gurdon showed that adult frogs could be generated from the nuclei of single somatic cells by nuclear transfer. In 2006, Yamanaka's group showed that somatic fibroblasts could be induced to ESC-like cells, called iPSCs, by four transcription factors including Oct4, Sox2, Klf4 and c-Myc. These iPSCs have the ability to generate all three lineages, endodermal, mesodermal and ectodermal cells [50]. Later, several groups using similar strategy successfully generated iPSCs from patients. Patient-derived iPSCs have been used to study the pathological mechanisms and drug testing [51, 52]. The principle of differentiated cells regaining pluripotency and conversion of one cell type into another not only let us re-think about the fundamental principles of development but also allow us re-consider autologous cell replacement therapy. Induced PSCs have been used for peripheral nerve repair. Uemura et al. transplanted iPSCs-derived neurospheres-seeded sponge polymer composed of 50% PLA and 50% PCL to 5 mm sciatic nerve gap. The recovery of motor and sensory function can be observed as early 4 weeks. Twelve weeks after transplantation, histological evaluation showed that iPSCs differentiated into GFAP-and S100-positive Schwann cells and Tuj1-and neurofilament-neuronal cells [53]. Wang et al. transplanted iPSCs-derived neural crest stem cells (NCSCs)-loaded electrospinning nanofibrous nerve conduits composed of 70% PLA and 30% PCL to 6 mm sciatic nerve gap. Transplanted NCSCs were able to promote regeneration of peripheral nerves. But they did not observed neuronal differentiation of NCSCs [54].

In 2010, Dr. Wernig's group used a cocktail of transcription factors, Ascl1, Brn2 and Myt1l, successfully convert fibroblasts into functional neurons, named induced neurons [55]. Interestingly, induced neurons have been generated from fibroblasts of patients [56]. Furthermore, induced dopaminergic neurons could integrate into host brain after transplantation. Recently, several group used similar techniques to successfully generate induced NSPCs [57, 58]. Induced NSPCs have similar properties with NSPCs isolated from brain tissue. More interestingly, two groups generated iOPCs from somatic fibroblasts. After transplantation, iOPCs could differentiate into oligodendrocytes and form myelin sheath. So far, there are no reports showed that iOPCs have been used to treat peripheral nerve injury [14, 15].

3. Biodegradable polymers as nerve guidance conduits for nerve repair and regeneration

The current gold standard for peripheral nerve repair is nerve autografting. However, this approach is associated with a number of clinical complications, in particular, donor site morbidity, limited availability, and nerve site mismatch and neuroma formation at the donor site [59]. Artificial nerve guidance conduits have been developed for bridging the gap. This strategy has been widely accepted for basic research and clinical applications. In general, a nerve guidance conduit for reconnecting the two nerve stumps (i.e., the proximal and distal nerve stumps) contains an appropriate substrate with longitudinal orientation guidance to direct axons to find their targets. To date, the majority of nerve guidance conduits developed is composed of biodegradable polymers. A few of them are FDA-approved and commercially available. These nerve conduits have achieved considerable success in treatment of gap defects with the distances up to 20-25 mm. To further improve on the regeneration capacity of nerve conduits, a number of strategies have been actively pursued, in particular, the combination of nerve conduits with stem cell technologies [60]. In this review, we discuss the key design parameters of nerve conduits, including materials, fabrications methods, and incorporation of bioactive molecules, which play critical roles in governing the cellular fate of the stem cells cultivated in the nerve guidance conduits. Through analyzing and summarizing these experimental results, our goal is to provide insights into the future design of nerve conduits with much improved therapeutic efficacy.

3.1. Materials consideration for use in nerve guidance conduits

Materials selection is critical to the performance of fabricated nerve guidance conduits. Ideally, materials to be employed for nerve repair need to fulfill the following requirements.

1. They must be biocompatibility for supporting cell growth, differentiation and function.
2. They must be able to provide appropriate surface and mechanical properties that mimic nerve tissue.
3. They must be immunologically inert.
4. They must be biodegradable or bio-absorbable.
5. They must be able to be sterilized.
6. They must be readily fabricated into the desired configurations of conduits.
7. Their production must be amenable to industrial scale-up.

3.2. Materials used in nerve guidance conduits

In the past decades, a broad range of materials have been explored for preparation of nerve guidance conduits for peripheral nerve repair. They can be broadly divided into two categories: naturally-derived biopolymers and synthetic polymers. As can be seen from the following discussion, these materials can address the aforementioned materials considerations to various degrees.

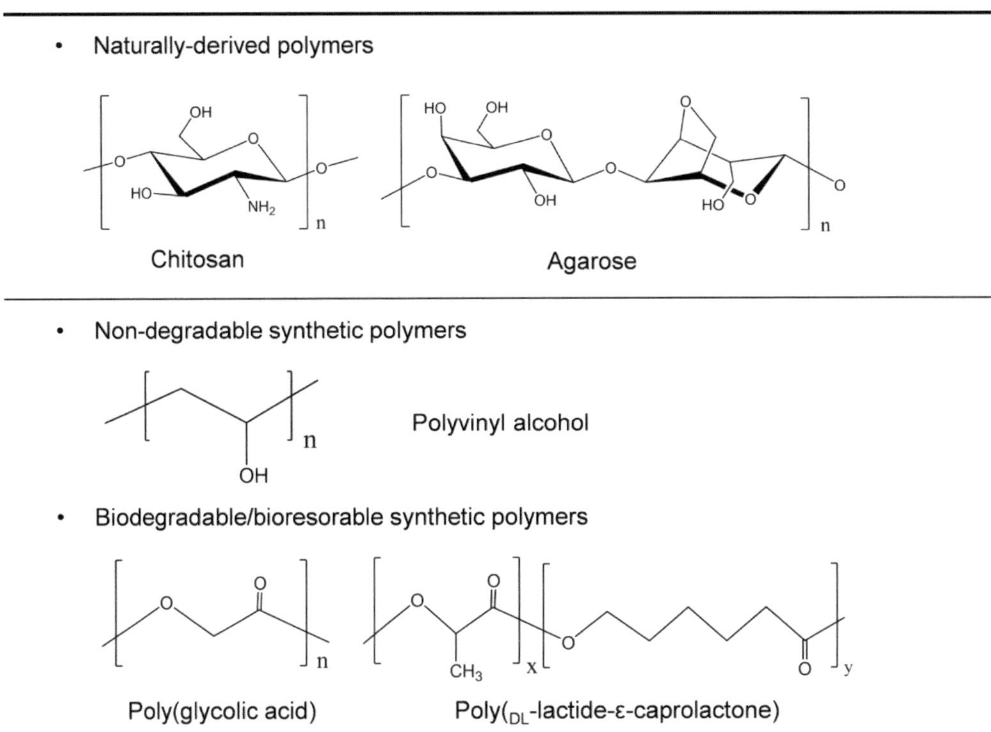

Table 1. Examples of naturally-derived and synthetic polymers used in nerve guidance conduits.

Naturally-derived biopolymers. Examples of the naturally-derived biopolymers for nerve guidance conduits include type I collagen, fibronectin, fibrin glue, gelatin, hyaluronic acid, alginate, chitosan, agarose and silk fibroin *etc*. These materials are typically hydrophilic, and form hydrogel-based matrices by either physical or chemical crosslinking. It is also worth noting two types of multicomponent matrices, such as Matrigel™ and decellularized nerve allografts or xenografts. Matrigel™ is a commercial extracellular matrix (ECM) extract from tissue cultured mouse sarcoma cell lines. It gels *in situ* at room temperature. Matrigel™ contains laminin, heparin sulphate, type IV collagen, entactin, nidogen and growth factors. Decellularized nerve allografts or xenografts are matrices that preserve the inherent structural characteristics of nerve. For instance, Avance® Nerve Graft is a commercially available decellularized allograft processed from human peripheral nerve tissue, which has been used clinically for reconstruction of periphery nerve gaps with positive results [61].

Naturally-derived biopolymers have demonstrated a number of advantages in the applications of nerve repair. They are in general biocompatible, and biodegradable or bioresorbable. They form hydrogels with structural and mechanical properties similar to nerve tissue. In addition, some of the biopolymers contain cell adhesion moieties, such as type I collagen and fibronectin, which encourages neuronal attachment and outgrowth. However, immunogenicity as a result of animal resources remains a major concern over their practical applications. In the case of Matrigel™, its origin in mouse sarcoma cells plagues its use in clinical applications. Naturally-derived polymers are also known for batch-to-batch variations, and lack of flexibility in terms of structural engineering to modulate the physiochemical properties and degradation kinetics of materials.

Synthetic polymers. Compared to naturally-derived biopolymers, synthetic polymers offer greater flexibility in modulating the physical properties of materials through engineering the polymer composition such as (co)monomer structure and side chain chemistry, and polymer molecular weight. They are also much more amenable to various technologies for materials fabrication. On the other hand, it is noted that the majority of synthetic polymers used in nerve repair lack biocompatibility and bioactivity, which limits cell attachment, growth and differentiation. Viable approaches to overcome the limit involve compositing synthetic polymers with bioactive molecules, and surface functionalization of nerve conduits with bioactive molecules. These approaches will be discussed in detail in Section 3.3.

At the early stage of nerve guide development, several non-biodegradable polymers have been explored to produce nerve guides, including, for example, silicone rubber, polyvinyl alcohol (PVA), polyethylene glycol (PEG), polyN-2-hydroxypropyl-methacrylamide (pHPMA) and poly2-hydroxyethyl methacrylate (pHEMA). A major disadvantage of non-biodegradable nerve conduits is that a second surgery is required to remove the conduits as their chronic presence impede nerves remodeling. In addition, studies have shown that the chronic presence of non-biodegradable conduits led to the inflammatory reactions and scar tissue formation, which ultimately inhibit functional nerve recovery [62, 63].

Amongst the synthetic polymers explored for nerve guidance conduits, a range of biodegradable aliphatic polyesters have attracted most attention. These include, for example, poly(glycolic acid) (PGA), polylactide (PLA), poly(lactic-co-glycolic acid) (PLGA), poly-ε-caprolactone (PCL) and poly-3-hydroxybutyrate (PHB) *etc*. These material typically produce relatively rigid scaffolds with hydrophobic surface. Their degradation is predominantly medicated by hydrolysis of their ester linkages in physiological conditions. The degradation rate is dependent on the polymer structure, molecular weight, or its crystallinity. For instance, among PGA, PLA and PCL of similar molecular weights, the degradation rate is PGA > PLA > PCL, as a result of increased hydrophobicity [64]. This provides a basis for tailoring the polymer degradation kinetics by varying the structure and ratio of the monomers used for polymerization. A good example is PLGA, whose degradation rate depends on the ratio of lactide to glycolide used for the polymerization, i.e., higher content of glycolide units leads to faster degradation rate, with an exception of 50:50 monomers' ratio that gives rise to the fastest degradation rate [65]. Synthetic biodegradable polymers have been widely used to fabricate different kinds of films or scaffolds to support cell growth in vitro and in vivo.

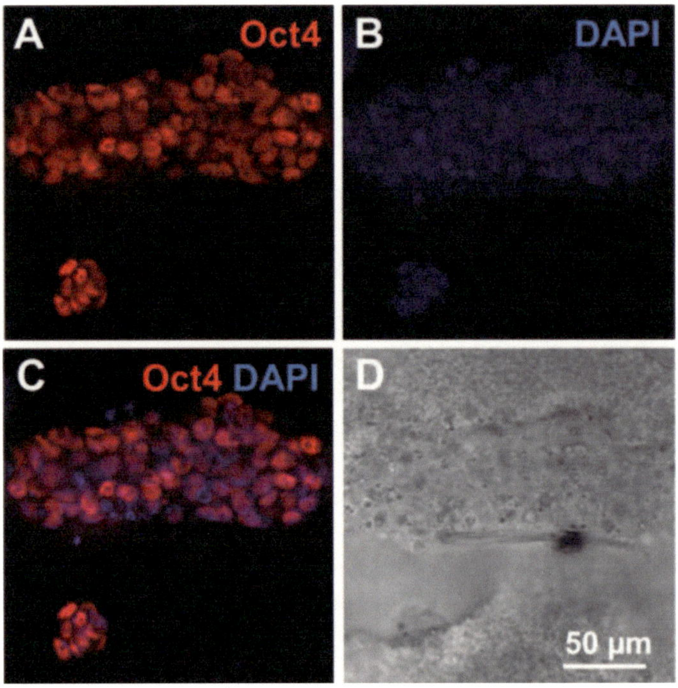

Figure 1. Immunocytochemistry assay showed proliferation of embryonic stem cells in 3D cellulosic hydrogel scaffolds after 2 days plating. A. Oct4 showed the pluripotency of embryonic stem cells. B. The total cells were showed by DAPI. C. A merges with B. D. DIC imaging showed proliferation of embryonic stem cells in 3D hydrogel scaffolds.

Despite the substantial research activities in engineering of polymer biodegradability, development of neural matrices with desired degradation kinetics in the course of nerve regeneration still remains a key challenge. This is due to the inherent complexity of *in vivo*-degradation, which necessitates multidisciplinary efforts to bring together materials scientists, biologists and clinicians to tackle this challenge. Ideally, a nerve conduit should provide adequate mechanical support and protection to facilitate axonal regeneration across the nerve gap, while undergoing degradation with the kinetics matching the rate of nerve regeneration, in order to make way for the regenerating nerve. Development of appropriate *in vitro*-models that can simulate *in vivo* degradation of nerve conduits may help to speed up the problem-solving process and facilitate the delivery of nerve conduits with desired *in vivo*-degradation profiles to meet specific needs in clinical applications.

3.3. Key considerations in nerve conduit design

In its simplest form, a nerve conduit takes the form of a hollow tube for bridging nerve gap defects. A nerve conduit implant needs to satisfy a set of basic requirements in both material and technical aspects, including biocompatibility, non-immunogenicity, biodegradability/ bioabsorbability, mechanical integrity with nerve tissue, and ease of sterilization and fabrication into the desired dimensions. These considerations underpin the development of a variety of single lumen hollow conduits that have shown positive effects in treatment of short defects (< 20-25 mm). These conduits differed mostly in the structure and composition of the polymers used for conduit fabrication. Some of them have already gained approval in clinical applications, which will be discussed in detail in Section 3.4.

To improve the therapeutic efficacy in treatment of large nerve defects, a number of biomimetic strategies have recently been adopted to modify the existing design of nerve conduits, by providing bio-regulative cues to better mimic nerve tissue at various levels (anatomic, physical or structural). The ongoing research activities in this area have been catalyzed by our increasing understanding of *in vivo*-behavior of nerve conduits, as well as the advances in material fabrication and in tissue engineering. Some of the key strategies will be briefly reviewed here.

Topographic guidance. Longitudinally-oriented topographic cues have been introduced to nerve conduit design, with a view to promoting the growth and orientation of regenerating axons. This can be microfibers, multichannels, or 3D matrix fillers with longitudinally-oriented architectures, as is illustrated in Figure 2. For instance, Kim *et al* reported on polysulfone conduits containing uniaxially electrospun nanofibers of poly(acrylonitrile-co-methylacrylate) [66]. Incorporation of such aligned sub-micron topographic cues was shown to significantly promote both sensory and motor nerve regeneration across a 17 mm peripheral nerve gap in a rodent model, without the delivery of any exogenous neuro-simulative agents (e.g., neurotrophic factors and extracellular matrix proteins). Nerve conduits with aligned multichannels can be prepared using an injection-molding technique, the same technique as used for fabrication of single lumen nerve conduits.

ECM proteins, such as laminin, fibronectin, and collagen, are good candidates of intraluminal fillers for nerve conduits as they promote axonal extension. Technologies are required in order to produce aligned intraluminal structures of these proteins. An early study by Dubey *et al* employed magnetic fields to align the collagen gels in Teflon tubes, which gave rise to enhanced neurite elongation from dorsal root ganglia explants [67]. Matsumoto *et al* developed nerve conduits of PGA that were further coated with collagen and internally filled with longitudinally-oriented, laminin-coated collagen microfibers [68]. Axonal regeneration over an 80 mm gap of canine peroneal nerves was demonstrated. Alternatively, using a special freezing process, 3D matrixes collagen with longitudinally oriented pores can be prepared, which may find applications as aligned intraluminal fillers [69].

Mechanical compliance. Mechanical compliance of nerve conduits is an important consideration in conduit design. It plays a key role in directing cellular and tissue response to implanted nerve conduits, which could affect the performance of the nerve conduits. In addition, studies have shown a critical role of material mechanical properties in directing stem cell differentiation [70, 71]. Nerve tissues are soft and highly hydrated, while nerve conduits prepared from synthetic polymers, such as polyesters, are often rigid and hydrophobic. Strategies have been sought to develop hydride conduits that integrate synthetic polymer with naturally-derived biopolymers into various configurations to unify the advantages of both types of materials.

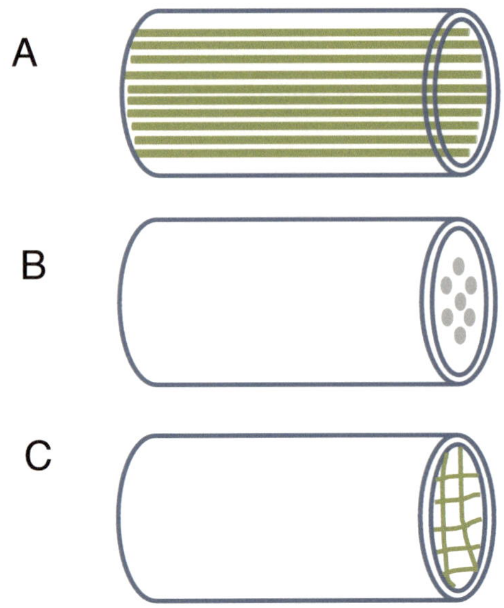

Figure 2. Schematic illustration of the nerve conduits modified with longitudinally oriented topographic cues. (S) nerve conduit with aligned nanofibres or microfibers; (B) nerve conduit containing multichannels; (C) nerve conduit containing 3D scaffolds with longitudinally oriented pores.

Examples of hydride conduits are those made of composites of naturally-derived biopolymers and synthetic polymers, or those incorporated with a soft hydrogels of ECM proteins to provide the matrix for axonal growth and regeneration [72, 73]. As previously discussed, those ECM hydrogel filaments need to be aligned to enable optimal nerve regeneration.

Surface bioengineering. This strategy involves coating or surface modification of nerve conduits with neurostimulatory molecules, in particular cell adhesion molecules, to promote cell adhesion, proliferation and differentiation, thereby improving the regeneration capacity of nerve conduits. The cell adhesion motifs that have been explored for nerve conduits include collagen, laminin, lamnin fragment peptides, fibronectin and Arg-Gly-Asp (RGD). For instance, laminin has often been used for coating of nerve conduits, and has been shown to enhance Schwann proliferation and migration, and neurite outgrowth. RGD-modified PCL nanofibers using a polyether diisocynate was shown to promote faster Schwann cell migration and axonal growth [74].

Growth factor delivery. As discussed previously, neutrotrophic factors play a critical role in promoting neuronal survival and differentiation. The capability to *in situ* deliver neurotrohic factors is now becoming an essential feature of next generation of nerve conduits. A popular approach is embedding growth factors in a hydrogel matrix that serves as intraluminal filler. Release of the entrapped growth factors can be diffusion-controlled and/or degradation-controlled, which is subject to the nature of the matrix-growth factor interactions. Increasing the crosslinking density of the hydrogel matrix can produce stiffer matrix, which may lead to more retarded release of the entrapped growth factor. However, this approach may have limitations as the level of stiffness of a hydrogel matrix should not pose any hindrances to the axonal growth across the lumen. Inclusion of heparin as a component of hydrogel matrix has shown a viable approach to modulate heparin-binding growth factors delivery. For instance, a heparin-containing hydrogel matrix was reported, comprising of fibrin with a high excess of immobilized heparin-binding peptides, heparin and neurotrophins, such as beta-nerve growth factor (β-NGF), brain-derived neurotrophic factor (BDNF) and neurotrophin-3 (NT-3) [75]. The heparin bound to both the immobilized peptides, and neurotrophins, which was responsible for slow diffusion release of β-NGF, BDNF or NT-3. Enhanced neurite extension was demonstrated in these heparin-containing matrices, but not in the matrices containing only fibrin and neurotrophins. A recent study also showed that the heparin-containing fibrin matrices with NGF resulted in a high level of sciatic nerve regeneration as compared to the control groups [76].

Alternatively, growth factors can be introduced into nerve conduits either as a component of coating or being embedded directly in the wall. For example, NGF-containing microspheres of PLGA were formulated with an aqueous solution of poly(2-hydroxyethyl methacrylate), and coated on the inside of pre-formed nerve conduits prepared from poly(2-hydroxyethyl methacrylate-*co*-methyl methacrylate) [77]. The microsphere-coated nerve conduits showed more sustained release of NGF for > 28 days, compared to those coated with poly(2-hydroxyethyl methacrylate) and NGF, though no *in vivo* studies were reported. Nerve conduits with the capacity of co-delivery of synergistically acting glial cell-line derived neurotrophic factors (GDNF) and NGF were reported by Madduri et al [78-80]. GDNF and NGF were loaded into the nerve conduits of collagen, and dried and coated with PLGA in ethyl acetate. *In vitro* studies showed that the combination of GDNF and NGF exerted a synergistic effect on the axonal elongation, axonal branching and growth kinetics. Compared to the conduits releasing GDNF alone, enhanced early nerve regeneration in a 10 mm rat sciatic nerve gap model was also demonstrated for the conduits with co-delivery of GDNF and NGF.

3.4. Commercially available nerve guidance conduits

Table 2 summarizes the nerve guidance conduits that have been approved by the US Food and Drug Administration (FDA), and/or the European Union with a Conformité Européenne certification (CE) for clinical applications. They are all in the configuration of single-lumen tube. NeuraGen™, NeuroMatrix™ and NeuroFlex™ are derived from type I collagen, which is a major component of ECM. AxoGuard™ is made of ECM materials derived from porcine small intestine. It contains almost intact ECM, including cell adhesion proteins, growth factors, glycosaminoglycans and proteoglycans *etc*. SaluBridge™ and SaluTunnel™ are non-resorbable conduits, and are prepared from PVA hydrogel. Neurotube® is a PGA-based, woven tubular device, with high porosity to provide an oxygen-rich environment for the regenerating nerve [60]. Neurolac® is the only FDA approved transparent conduit based on synthetic biodegradable polyesters. It is noted that the fabrication of these nerve guides do not involve any biofunctionalization with or incorporation of bioactive molecules. These nerve conduits thus meet only the basic requirement of conduits, by providing physical guidance cues via conduit morphology to direct axonal regeneration.

Product name	Materials	Degradation	Some key issues
SaluBridge™, SaluTunnel™ (from SaluMedica)	PVA	Non-resorbable	Non-resorbability
Neuroflex™, NeuroMatrixTM (from Collagen Matrix Inc.)	Type I bovine collagen	4-8 months	Risk of adverse immune response
NeuraGen® (from Integra life Science)	Type I bovine collagen	36-48 months	Risk of adverse immune response
AxoGuardTM (from Cook Biotech)	Porcine small intestinal submucosa (SIS)	3 months	Risk of adverse immune response; Risk of infectious disease transmission
Neurotube® (from Synovis® Micro)	PGA	3 months	Rapid loss of mechanical properties; acid degradation products
Neurolac® (Polyganics B.V.)	Poly($_{DL}$-lactide-ε-caprolactone)	16 months	Rigidity and inflexibility; foreign body reactions; polymer fragments

Table 2. Clinically-approved nerve guidance conduits (adapt with permission from [60, 81-83]).

3.5. Next generation of nerve guidance conduits in tandem with stem cell delivery

Stem cells have been used with nerve conduits or 3D scaffolds to obtain the maximum efficient therapeutic effects of nerve repair. Stem cells-loaded nerve conduits and 3D scaffolds have much higher potential for nerve repair compared with nerve conduits or 3D scaffolds alone. For example, Park et al. cultured NSCs in PGA scaffolds for 4 days before transplantation. And then NSC-PGA complexes were transplanted into the infarction cavity of the brains in the mouse model of hypoxia ischemic injury by glass micropipettes. The results showed that PGA provides a good support the survival and neuronal differentiation of transplanted NSCs. Transplanted NSCs differentiates into neurons in the infarct area. Antegrade and retrograde tract tracing showed that transplanted projects the axons to internal and external capsule and the contralateral hemisphere through corpus callosum. Animal behavioral function has not been tested in this report [84]. Liu et al. transplanted ADSCs-loaded biodegradable GGT nerve conduits containing genipin crosslinked gelatin annexed with tricalcium phosphate (TCP) ceramic particles into 10 mm gap in the sciatic nerve after injury. They found that ADSCs could differentiate into neuron-like cells in the GGT nerve conduits and ADSCs-loaded biodegradable GGT nerve conduits significantly increased sciatic function index and functional recovery [85]. Furthermore, BDNF or GDNF-transfected NSCs-seeded PLA microporous nerve conduits were transplanted into sciatic nerve gap after injury. The conduits seeded with GDNF-and BDNF-transfected NSCs significantly increased the degree of myelination and the size of regenerated tissue compared with those seeded with the nontransfected NSCs. The greatest number of blood vessels was found in the animals transplanted with GDNF-transfected NSCs-seeded PLA microporous nerve conduits. The functional recovery was significantly improved for BDNF or GDNF-transfected NSCs-seeded conduits assessed by the functional gait and electrophysiology [86].

PCL conduits filled with bone marrow-derived MSCs were tested for repair of transected sciatic nerves in mice [87]. Use of the MSCs grafted conduits was shown to significantly improve the survival of sensory neurons and motor function, and restore gastrocnemius muscle function in mice. In a separate study, MSCs were loaded to silk fibroin-based conduits that were filled with oriented silk fibroin filaments, which were tested for bridging a 10 mm-long gap in rat sciatic nerve [88]. At the early weeks after nerve grafting, the grafted MSCs were shown to enhance the gene expression of several growth factors, such as BDNF, bFGF, and ciliary neurotrophic factor, and S100, a marker of Schwann cells. These were arguably responsible for accelerated axonal elongation at 4 weeks, and an improved outcome in sciatic nerve regeneration and functional recovery at weeks 12 in the groups treated with MSC-grafted conduits, when compared to those treated with acellular conduits. The nerve regeneration efficacy of the MSC-grafted conduits was shown to approach that of autologous nerve grafts.

4. Nerve conduits in clinical applications

In the past decades, autogenous and polymer-based nerve conduits have been used for nerve repair in clinic. Both of them have showed positive clinical outcomes in the patients.

4.1. Biological autogenous nerve conduits in clinical applications

Previous clinical studies show that autogenous vein conduits are able to repair nerve injury in patients. A retrospective clinical study evaluated 22 digital nerve repairs in the finger using autogenous vein conduits, and reported that two-point discrimination for 11 acute digital nerve repairs with vein grafts and poor results for delayed digital nerve repair [89]. In 1990, Chiu and Strauch compared autogenous vein grafts with conventional nerve graft in 22 patients. They demonstrated that autogenous vein grafts were as efficient as conventional nerve grafts to bridge a small nerve gap (≤3 cm). But this study did show how long these patients were operated after nerve injury [90]. Similar clinical study was performed by Laveaux et al., and reported that vein grafts is less efficient than nerve grafts in delayed nerve repair and vein grafts produce similar good results in emergency cases [91]. A long-term sensory evaluation of nerve repair was performed by Lee and Shieh, and reported that vein conduit grafts could produce excellent sensory recovery [92]. In 2001, Pogrel and Maghen used autogenous vein grafts to repair continuity defects, ranged from 2 to 14 mm, of the inferior alveolar nerves (n=6) and lingual nerves (n=10). All the patients received grafts between 4 and 10 months after injury. They found that vein grafts can form a physiological conduit for nerve regeneration and are more successful with short gaps [93]. More recent interesting study implanted male vein grafts to femoral nerve injury of female rats and found that male vein cells could integrate into female injured nerve and participate in remyelination and nerve regeneration [94]. In 2012, Liard et al. evaluated that adult neural stem cells-loaded autogenous vein grafts to reconstruct nerve gaps in pig model and demonstrated that neural stem cells transplantation increased 2',3'-cyclic nucleotide 3'-phosphodiesterase (CNPase) expression and promoted functional recovery of stimulodetection [95].

Autogenous muscle grafts are another option to provide a scaffold for the nerve fiber to grow. In 1988, Norris et al. used frozen and thawed to denature skeletal muscle and transplanted skeletal muscle to injured digital nerves in 8 patients. 7 out of 8 showed an excellent level of recovery, MRC sensory clinical score $S3^{+}$[96]. In 2008, Pereira et al. treated 38 patients with leprosy by skeletal muscle autografts ranging between 2.5 cm and 14 cm length. The clinical results showed that sensory recovery was noted in 89% patients and 80 % of ulcers caused by posterior tibial nerve damage were healed [97]. Furthermore, to increase clinical effects, the vein conduits filled with muscle are also used to bridge peripheral nerve gaps. The basic idea is that vein could provide regeneration guidance and muscle serves supporter to avoid vein collapse. In 1993, Brunelli et al. reported that vein plus muscle grafts could have similar functional recovery to those found in traditional nerve grafts. More interestingly, axon number in vein plus muscle grafts group is significantly higher than that of traditional nerve grafts group [98]. In 2000, similar work was done by Battiston et al. 21 patients suffered nerve defects of 5-60 mm were treated with vein filled with skeletal muscle. 85% of patients showed good clinical results [99].

Although vein and muscle grafts are more available than nerve grafts, isolating vein and muscle grafts also need second operation. To overcome this critical clinical problem, scientists have developed different synthetic polymer-based nerve conduits for nerve repair.

4.2. Synthetic polymer-based nerve conduits in clinical applications

In 1998, Sanda Stanec and Zdenko Stanec used non-absorable polytetrafluoroethylene (ePTFE) tube to reconstruct nerve defects between 1.5 to 6 cm length, and demonstrated that 78.6% patients suffered 1.5 to 4 cm length nerve defects had functional motor and sensory recovery, but only 13.3% patients suffered 4.1 to 6 cm length defects had similar recovery [100]. Due to non-absorable nerve conduits need secondary surgery to remove them, most of synthetic nerve conduits are made of biodegradable materials. In 2009, Rosson et al. evaluated 6 patients with short-gap motor nerve injuries treated with bioabsorbable conduit, the Neurotube™, and observed that all patients had some return of motor function. It demonstrated that motor nerves with short-gap injuries could regenerate cross this conduit [101]. In 2011, Rinker and Liau compared the clinical output of woven polyglycolic acid and autogenous vein conduits for reconstruction of digital nerve gaps and reported that sensory recovery after digital nerve reconstruction with autogenous vein conduit was similar to that using polyglycolic acid conduit and similar cost profile and less postoperative complications were observed in both of them [102]. Taras et al. reconstructed 22 isolated digital nerve lacerations in 19 patients with a bioabsorable collagen conduit, and showed that 13 out of 22 achieved excellent results, 3 of 22 obtained good results, and there were no poor results [103]. A retrospective study of 10 cases was performed by Thomsen et al. in 2010. All patients were operated on for painful neuroma and underwent repair with collagen conduits (Revolnerv®, Orthomed). Fifty percent patients had excellent or good results at static two-point discrimination testing [104].

Although nerve conduits are commercially available, their clinical application is far satisfied. Limitations of nerve conduits in peripheral nerve repairs were reported by Moore et al, in 2009. In this 4 cases report, 3 patients were treated with type I collagen nerve conduit (NeuraGen, Integra NeuroSciences) and 1 patient was treated with polyglycolic acid nerve conduit (GEM Neurotube, Synovis, Birmingham, AL, USA). There were no clinical effects in these patients. Side effects were reported by some patients [105]. In 2010, Wangensteen and Kalliainen reviewed 96 patients' clinical data, who received type I collagen nerve conduit (NeuraGen, Integra NeuroSciences) for nerve repair. Only 35-45% patients had sensory recovery [106].

5. Perspectives

Recent progress of biodegradable materials and stem cells provides more options for nerve regeneration. Neural tissue engineering is a new thing but has been widely used for nerve regeneration in basic research and clinical application. Although the research of peripheral nerve repair has started many years ago, functional recovery is still unsatisfied. The functional recovery largely depends on nerve gap, the location of injured nerve, patients' age and methods of treatment chosen. From the literatures, there are limited choices for nerve regeneration: (1) For tiny nerve gap, microsurgery joining the distal and proximal stumps of the damaged nerves should be first choice; (2) For small nerve gap (≤2-3 cm), autogenous nerve or vein grafts and acellular nerve conduits can be used for nerve repair; (3) For larger nerve gap (≥3 cm), just nerve conduits are not enough to support nerve regeneration. The studies of animal trials showed that combination of nerve conduits with supporting cells could be best choice to obtain maximum extent functional recovery. However, most popular animal model for studying peripheral nerve regeneration is rat sciatic nerve injury model. Rat is a small animal compared with human. It is impossible to expect that the similar functional recovery would be obtained in the peripheral nerve injury patients with similar treatment done in rat animal model. More works need to be done with large animals, such as monkey, to optimize the approaches. Furthermore, personal medicine for cell therapy needs patients-derived cells. Experience with induced pluripotent stem cells, induced neural stem cells, and induced neurons make it possible to generate large quantity of patients-derived cells for clinical application. Practically, multiple-disciplinary approaches should be combined together to generate optimal clinical recovery for patients with peripheral nerve injury.

Acknowledgements

This work was supported by NIH Grant DA025143, OD008761 and NARSAD Young Investigator Award (Dr. Zhang).

Author details

Haigang Gu[1], Zhilian Yue[2] and Qi Zhang[1]

1 Department of Pharmacology, Vanderbilt University School of Medicine, Nashville, USA

2 Intelligent Polymer Research Institute, AIIM Facility, Innovation Campus, University of Wollongong, Australia

References

[1] Ludwin SK: Remyelination in the central nervous system and the peripheral nervous system. Advances in neurology 1988, 47:215-254.

[2] Park DH, Eve DJ, Chung YG, Sanberg PR: Regenerative medicine for neurological disorders. ScientificWorldJournal 2010, 10:470-489.

[3] Jiang B, Zhang P, Zhang D, Fu Z, Yin X, Zhang H: Study on small gap sleeve bridging peripheral nerve injury. Artificial cells, blood substitutes, and immobilization biotechnology 2006, 34(1):55-74.

[4] Radtke C, Wewetzer K, Reimers K, Vogt PM: Transplantation of olfactory ensheathing cells as adjunct cell therapy for peripheral nerve injury. Cell Transplant 2011, 20(2):145-152.

[5] Cote MP, Amin AA, Tom VJ, Houle JD: Peripheral nerve grafts support regeneration after spinal cord injury. Neurotherapeutics 2011, 8(2):294-303.

[6] Lineaweaver W: Immediate nerve grafts to a median nerve injury in a 7-year-old boy: 5-year follow-up. Journal of the Mississippi State Medical Association 2013, 54(9): 252-254.

[7] Gage FH, Temple S: Neural stem cells: generating and regenerating the brain. Neuron 2013, 80(3):588-601.

[8] Krencik R, Weick JP, Liu Y, Zhang ZJ, Zhang SC: Specification of transplantable astroglial subtypes from human pluripotent stem cells. Nature biotechnology 2011, 29(6):528-534.

[9] Orlacchio A, Bernardi G, Orlacchio A, Martino S: Stem cells: an overview of the current status of therapies for central and peripheral nervous system diseases. Current medicinal chemistry 2010, 17(7):595-608.

[10] Gu H: Stem Cell-Derived Neurons for the Treatment of Neurodegenerative Diseases. Clinic Pharmacol Biopharmaceut 2013, 2(111):doi:10.4172/2167-4065X.1000111.

[11] Gu H: Modeling and Therapeutic Strategies of Pluripotent Stem Cells for Alzheimer's Disease. J Stem Cell Res Ther 2013, 3(e115):doi:10.4172/2157-7633.1000e4115.

[12] Gu H: Using induced pluripotent stem cells to model neurodegenerative disease. J Anc Dis Prev Rem 2013, 1:e101.

[13] Xuan AG, Long DH, Gu HG, Yang DD, Hong LP, Leng SL: BDNF improves the effects of neural stem cells on the rat model of Alzheimer's disease with unilateral lesion of fimbria-fornix. Neurosci Lett 2008, 440(3):331-335.

[14] Yang N, Zuchero JB, Ahlenius H, Marro S, Ng YH, Vierbuchen T, Hawkins JS, Geissler R, Barres BA, Wernig M: Generation of oligodendroglial cells by direct lineage conversion. Nature biotechnology 2013, 31(5):434-439.

[15] Najm FJ, Lager AM, Zaremba A, Wyatt K, Caprariello AV, Factor DC, Karl RT, Maeda T, Miller RH, Tesar PJ: Transcription factor-mediated reprogramming of fibroblasts to expandable, myelinogenic oligodendrocyte progenitor cells. Nature biotechnology 2013, 31(5):426-433.

[16] Zarbakhsh S, Bakhtiyari M, Faghihi A, Joghataei MT, Mehdizadeh M, Khoei S, Mansouri K, Yousefi B, Pirhajati V, Moradi F *et al*: The effects of schwann and bone marrow stromal stem cells on sciatic nerve injury in rat: a comparison of functional recovery. Cell journal 2012, 14(1):39-46.

[17] Frostick SP, Yin Q, Kemp GJ: Schwann cells, neurotrophic factors, and peripheral nerve regeneration. Microsurgery 1998, 18(7):397-405.

[18] Lehmann HC, Hoke A: Schwann cells as a therapeutic target for peripheral neuropathies. CNS Neurol Disord Drug Targets 2010, 9(6):801-806.

[19] Salzer JL, Bunge RP: Studies of Schwann cell proliferation. I. An analysis in tissue culture of proliferation during development, Wallerian degeneration, and direct injury. The Journal of cell biology 1980, 84(3):739-752.

[20] Morrissey TK, Kleitman N, Bunge RP: Isolation and functional characterization of Schwann cells derived from adult peripheral nerve. The Journal of neuroscience : the official journal of the Society for Neuroscience 1991, 11(8):2433-2442.

[21] Imaizumi T, Lankford KL, Kocsis JD: Transplantation of olfactory ensheathing cells or Schwann cells restores rapid and secure conduction across the transected spinal cord. Brain research 2000, 854(1-2):70-78.

[22] Guenard V, Kleitman N, Morrissey TK, Bunge RP, Aebischer P: Syngeneic Schwann cells derived from adult nerves seeded in semipermeable guidance channels enhance peripheral nerve regeneration. The Journal of neuroscience : the official journal of the Society for Neuroscience 1992, 12(9):3310-3320.

[23] Berrocal YA, Almeida VW, Gupta R, Levi AD: Transplantation of Schwann cells in a collagen tube for the repair of large, segmental peripheral nerve defects in rats. Journal of neurosurgery 2013, 119(3):720-732.

[24] Bianco P, Robey PG: Stem cells in tissue engineering. Nature 2001, 414(6859):118-121.

[25] Bianco P, Riminucci M, Gronthos S, Robey PG: Bone marrow stromal stem cells: nature, biology, and potential applications. Stem Cells 2001, 19(3):180-192.

[26] Vaananen HK: Mesenchymal stem cells. Annals of medicine 2005, 37(7):469-479.

[27] Uccelli A, Moretta L, Pistoia V: Mesenchymal stem cells in health and disease. Nature reviews Immunology 2008, 8(9):726-736.

[28] Bianco P, Cao X, Frenette PS, Mao JJ, Robey PG, Simmons PJ, Wang CY: The meaning, the sense and the significance: translating the science of mesenchymal stem cells into medicine. Nature medicine 2013, 19(1):35-42.

[29] Robey PG, Kuznetsov SA, Riminucci M, Bianco P: Bone marrow stromal cell assays: in vitro and in vivo. Methods Mol Biol 2014, 1130:279-293.

[30] Konno M, Hamabe A, Hasegawa S, Ogawa H, Fukusumi T, Nishikawa S, Ohta K, Kano Y, Ozaki M, Noguchi Y *et al*: Adipose-derived mesenchymal stem cells and regenerative medicine. Development, growth & differentiation 2013, 55(3):309-318.

[31] Nong Y, Zhang C, Wei L, Zhang Z, Cheng J, Wen L, Song Z: In situ investigation of allografted mouse HCN4 gene-transfected rat bone marrow mesenchymal stromal cells with the use of patch-clamp recording of ventricular slices. Cytotherapy 2013, 15(8):905-919.

[32] Ma K, Fox L, Shi G, Shen J, Liu Q, Pappas JD, Cheng J, Qu T: Generation of neural stem cell-like cells from bone marrow-derived human mesenchymal stem cells. Neurol Res 2011, 33(10):1083-1093.

[33] Gu H, Yue Z, Leong WS, Nugraha B, Tan LP: Control of in vitro neural differentiation of mesenchymal stem cells in 3D macroporous, cellulosic hydrogels. Regen Med 2010, 5(2):245-253.

[34] Usach V, Goitia B, Lavalle L, Martinez Vivot R, Setton-Avruj P: Bone marrow mononuclear cells migrate to the demyelinated sciatic nerve and transdifferentiate into Schwann cells after nerve injury: attempt at a peripheral nervous system intrinsic repair mechanism. J Neurosci Res 2011, 89(8):1203-1217.

[35] Lee JH, Chung WH, Kang EH, Chung DJ, Choi CB, Chang HS, Lee JH, Hwang SH, Han H, Choe BY *et al*: Schwann cell-like remyelination following transplantation of human umbilical cord blood (hUCB)-derived mesenchymal stem cells in dogs with acute spinal cord injury. Journal of the neurological sciences 2011, 300(1-2):86-96.

[36] Cho KJ, Trzaska KA, Greco SJ, McArdle J, Wang FS, Ye JH, Rameshwar P: Neurons derived from human mesenchymal stem cells show synaptic transmission and can be

induced to produce the neurotransmitter substance P by interleukin-1 alpha. Stem Cells 2005, 23(3):383-391.

[37] Fu YS, Cheng YC, Lin MY, Cheng H, Chu PM, Chou SC, Shih YH, Ko MH, Sung MS: Conversion of human umbilical cord mesenchymal stem cells in Wharton's jelly to dopaminergic neurons in vitro: potential therapeutic application for Parkinsonism. Stem Cells 2006, 24(1):115-124.

[38] Gage FH: Mammalian neural stem cells. Science 2000, 287(5457):1433-1438.

[39] Temple S: The development of neural stem cells. Nature 2001, 414(6859):112-117.

[40] Gu H, Yu SP, Gutekunst CA, Gross RE, Wei L: Inhibition of the Rho signaling pathway improves neurite outgrowth and neuronal differentiation of mouse neural stem cells. Int J Physiol Pathophysiol Pharmacol 2013, 5(1):11-20.

[41] Gu H, Long D, Leng S: Isolation cultivation of neural stem cells from basal forebrain of newborn rats. Anat Res 2003, 25(2):88-90; 98.

[42] Gu H, Long D, Li X, Leng S, Luo M: The survival and migration of neural stem cells in the basal forebrain of rats with Alzheimer's disese. Chin J Neuroanat 2008, 24(3): 245-250.

[43] Gu H, Long D, Li X, Zhang G, Luo M, Li J, Leng S: Proliferation and differentiation of neural stem cells from neonatal rat basal forebrain of newborn rats into neurons in different culture conditions. J Clin Rehabilit Tiss Eng Res 2008, 12(8):1445-1448.

[44] Gu H, Long D, Li X, Zhang G, Su T, Li J, Leng S: Effect of neural stem cells transplantation on parvalbumin-positive neurons of the basal forebrain and abilities of learning and memory in a rat model of senile dementia. J Clin Rehabilit Tiss Eng Res 2008, 12(12):2235-2239.

[45] Gu H, Long D, Song C, Li X: The effects of neural stem cell transplantation on cholinergic neurons of the basal forebrain and abilities of learning and memory of the rat model of Alzheimer's disease. Chin J Clin Anat 2009, 27(1):85-89.

[46] Wang JY, Liou AK, Ren ZH, Zhang L, Brown BN, Cui XT, Badylak SF, Cai YN, Guan YQ, Leak RK *et al*: Neurorestorative effect of urinary bladder matrix-mediated neural stem cell transplantation following traumatic brain injury in rats. CNS Neurol Disord Drug Targets 2013, 12(3):413-425.

[47] Kim H, Zahir T, Tator CH, Shoichet MS: Effects of dibutyryl cyclic-AMP on survival and neuronal differentiation of neural stem/progenitor cells transplanted into spinal cord injured rats. PLoS One 2011, 6(6):e21744.

[48] Kulbatski I, Mothe AJ, Parr AM, Kim H, Kang CE, Bozkurt G, Tator CH: Glial precursor cell transplantation therapy for neurotrauma and multiple sclerosis. Progress in histochemistry and cytochemistry 2008, 43(3):123-176.

[49] Olson HE, Rooney GE, Gross L, Nesbitt JJ, Galvin KE, Knight A, Chen B, Yaszemski MJ, Windebank AJ: Neural stem cell-and Schwann cell-loaded biodegradable poly-

mer scaffolds support axonal regeneration in the transected spinal cord. Tissue Eng Part A 2009, 15(7):1797-1805.

[50] Takahashi K, Yamanaka S: Induction of pluripotent stem cells from mouse embryonic and adult fibroblast cultures by defined factors. Cell 2006, 126(4):663-676.

[51] Agarwal S, Loh YH, McLoughlin EM, Huang J, Park IH, Miller JD, Huo H, Okuka M, Dos Reis RM, Loewer S *et al*: Telomere elongation in induced pluripotent stem cells from dyskeratosis congenita patients. Nature 2010, 464(7286):292-296.

[52] Dimos JT, Rodolfa KT, Niakan KK, Weisenthal LM, Mitsumoto H, Chung W, Croft GF, Saphier G, Leibel R, Goland R *et al*: Induced pluripotent stem cells generated from patients with ALS can be differentiated into motor neurons. Science 2008, 321(5893):1218-1221.

[53] Uemura T, Takamatsu K, Ikeda M, Okada M, Kazuki K, Ikada Y, Nakamura H: Transplantation of induced pluripotent stem cell-derived neurospheres for peripheral nerve repair. Biochem Biophys Res Commun 2012, 419(1):130-135.

[54] Wang A, Tang Z, Park IH, Zhu Y, Patel S, Daley GQ, Li S: Induced pluripotent stem cells for neural tissue engineering. Biomaterials 2011, 32(22):5023-5032.

[55] Vierbuchen T, Ostermeier A, Pang ZP, Kokubu Y, Sudhof TC, Wernig M: Direct conversion of fibroblasts to functional neurons by defined factors. Nature 2010, 463(7284):1035-1041.

[56] Qiang L, Fujita R, Yamashita T, Angulo S, Rhinn H, Rhee D, Doege C, Chau L, Aubry L, Vanti WB *et al*: Directed conversion of Alzheimer's disease patient skin fibroblasts into functional neurons. Cell 2011, 146(3):359-371.

[57] Han DW, Tapia N, Hermann A, Hemmer K, Hoing S, Arauzo-Bravo MJ, Zaehres H, Wu G, Frank S, Moritz S *et al*: Direct reprogramming of fibroblasts into neural stem cells by defined factors. Cell Stem Cell 2012, 10(4):465-472.

[58] Thier M, Worsdorfer P, Lakes YB, Gorris R, Herms S, Opitz T, Seiferling D, Quandel T, Hoffmann P, Nothen MM *et al*: Direct conversion of fibroblasts into stably expandable neural stem cells. Cell Stem Cell 2012, 10(4):473-479.

[59] Nectow AR, Marra KG, Kaplan DL: Biomaterials for the development of peripheral nerve guidance conduits. Tissue Eng Part B Rev 2012, 18(1):40-50.

[60] Kehoe S, Zhang XF, Boyd D: FDA approved guidance conduits and wraps for peripheral nerve injury: a review of materials and efficacy. Injury 2012, 43(5):553-572.

[61] Karabekmez FE, Duymaz A, Moran SL: Early clinical outcomes with the use of decellularized nerve allograft for repair of sensory defects within the hand. Hand 2009, 4(3):245-249.

[62] Braga-Silva J: The use of silicone tubing in the late repair of the median and ulnar nerves in the forearm. Journal of hand surgery 1999, 24(6):703-706.

[63] Merle M, Dellon AL, Campbell JN, Chang PS: Complications from silicon-polymer intubulation of nerves. Microsurgery 1989, 10(2):130-133.

[64] Jeong SI, Kim BS, Kang SW, Kwon JH, Lee YM, Kim SH, Kim YH: In vivo biocompatibilty and degradation behavior of elastic poly(L-lactide-co-epsilon-caprolactone) scaffolds. Biomaterials 2004, 25(28):5939-5946.

[65] Makadia HK, Siegel SJ: Poly Lactic-co-Glycolic Acid (PLGA) as Biodegradable Controlled Drug Delivery Carrier. Polymers 2011, 3(3):1377-1397.

[66] Kim YT, Haftel VK, Kumar S, Bellamkonda RV: The role of aligned polymer fiber-based constructs in the bridging of long peripheral nerve gaps. Biomaterials 2008, 29(21):3117-3127.

[67] Dubey N, Letourneau PC, Tranquillo RT: Guided neurite elongation and schwann cell invasion into magnetically aligned collagen in simulated peripheral nerve regeneration. Exp Neurol 1999, 158(2):338-350.

[68] Matsumoto K, Ohnishi K, Kiyotani T, Sekine T, Ueda H, Nakamura T, Endo K, Shimizu Y: Peripheral nerve regeneration across an 80-mm gap bridged by a polyglycolic acid (PGA)-collagen tube filled with laminin-coated collagen fibers: a histological and electrophysiological evaluation of regenerated nerves. Brain research 2000, 868(2):315-328.

[69] Mollers S, Heschel I, Damink LH, Schugner F, Deumens R, Muller B, Bozkurt A, Nava JG, Noth J, Brook GA: Cytocompatibility of a novel, longitudinally microstructured collagen scaffold intended for nerve tissue repair. Tissue Eng Part A 2009, 15(3):461-472.

[70] Discher DE, Janmey P, Wang YL: Tissue cells feel and respond to the stiffness of their substrate. Science 2005, 310(5751):1139-1143.

[71] Engler AJ, Sen S, Sweeney HL, Discher DE: Matrix elasticity directs stem cell lineage specification. Cell 2006, 126(4):677-689.

[72] Xie F, Li QF, Gu B, Liu K, Shen GX: In vitro and in vivo evaluation of a biodegradable chitosan-PLA composite peripheral nerve guide conduit material. Microsurgery 2008, 28(6):471-479.

[73] Hill PS, Apel PJ, Barnwell J, Smith T, Koman LA, Atala A, Van Dyke M: Repair of peripheral nerve defects in rabbits using keratin hydrogel scaffolds. Tissue Eng Part A 2011, 17(11-12):1499-1505.

[74] Bockelmann J, Klinkhammer K, von Holst A, Seiler N, Faissner A, Brook GA, Klee D, Mey J: Functionalization of electrospun poly(epsilon-caprolactone) fibers with the extracellular matrix-derived peptide GRGDS improves guidance of schwann cell migration and axonal growth. Tissue Eng Part A 2011, 17(3-4):475-486.

[75] Sakiyama-Elbert SE, Hubbell JA: Controlled release of nerve growth factor from a heparin-containing fibrin-based cell ingrowth matrix. J Control Release 2000, 69(1): 149-158.

[76] Wood MD, Hunter D, Mackinnon SE, Sakiyama-Elbert SE: Heparin-binding-affinity-based delivery systems releasing nerve growth factor enhance sciatic nerve regeneration. J Biomater Sci Polym Ed 2010, 21(6):771-787.

[77] Piotrowicz A, Shoichet MS: Nerve guidance channels as drug delivery vehicles. Biomaterials 2006, 27(9):2018-2027.

[78] Madduri S, di Summa P, Papaloizos M, Kalbermatten D, Gander B: Effect of controlled co-delivery of synergistic neurotrophic factors on early nerve regeneration in rats. Biomaterials 2010, 31(32):8402-8409.

[79] Madduri S, Feldman K, Tervoort T, Papaloizos M, Gander B: Collagen nerve conduits releasing the neurotrophic factors GDNF and NGF. J Control Release 2010, 143(2):168-174.

[80] Madduri S, Papaloizos M, Gander B: Synergistic effect of GDNF and NGF on axonal branching and elongation in vitro. Neuroscience research 2009, 65(1):88-97.

[81] Haug A: US Food and Drug Administration/Conformit Europe-approved absorbable nerve conduits for clinical repair of peripheral and cranial nerves. Annals of plastic surgery 2009, 62(6):710.

[82] Meek MF, Coert JH: US Food and Drug Administration /Conformit Europe-approved absorbable nerve conduits for clinical repair of peripheral and cranial nerves. Annals of plastic surgery 2008, 60(4):466-472.

[83] Meek MF, Coert JH: US Food and Drug Administration/Conformit Europe-approved absorbable nerve conduits for clinical repair of peripheral and cranial nerves. Annals of plastic surgery 2008, 60(1):110-116.

[84] Park KI, Teng YD, Snyder EY: The injured brain interacts reciprocally with neural stem cells supported by scaffolds to reconstitute lost tissue. Nature biotechnology 2002, 20(11):1111-1117.

[85] Liu BS, Yang YC, Shen CC: Regenerative effect of adipose tissue-derived stem cells transplantation using nerve conduit therapy on sciatic nerve injury in rats. J Tissue Eng Regen Med 2012.

[86] Fu KY, Dai LG, Chiu IM, Chen JR, Hsu SH: Sciatic nerve regeneration by microporous nerve conduits seeded with glial cell line-derived neurotrophic factor or brain-derived neurotrophic factor gene transfected neural stem cells. Artificial organs 2011, 35(4):363-372.

[87] Frattini F, Lopes FR, Almeida FM, Rodrigues RF, Boldrini LC, Tomaz MA, Baptista AF, Melo PA, Martinez AM: Mesenchymal stem cells in a polycaprolactone conduit

promote sciatic nerve regeneration and sensory neuron survival after nerve injury. Tissue Eng Part A 2012, 18(19-20):2030-2039.

[88] Yang Y, Yuan X, Ding F, Yao D, Gu Y, Liu J, Gu X: Repair of rat sciatic nerve gap by a silk fibroin-based scaffold added with bone marrow mesenchymal stem cells. Tissue Eng Part A 2011, 17(17-18):2231-2244.

[89] Walton RL, Brown RE, Matory WE, Jr., Borah GL, Dolph JL: Autogenous vein graft repair of digital nerve defects in the finger: a retrospective clinical study. Plastic and reconstructive surgery 1989, 84(6):944-949; discussion 950-942.

[90] Chiu DT, Strauch B: A prospective clinical evaluation of autogenous vein grafts used as a nerve conduit for distal sensory nerve defects of 3 cm or less. Plastic and reconstructive surgery 1990, 86(5):928-934.

[91] Laveaux C, Pauchot J, Obert L, Choserot V, Tropet Y: [Retrospective monocentric comparative evaluation by sifting of vein grafts versus nerve grafts in palmar digital nerves defects. Report of 32 cases]. Annales de chirurgie plastique et esthetique 2010, 55(1):19-34.

[92] Lee YH, Shieh SJ: Secondary nerve reconstruction using vein conduit grafts for neglected digital nerve injuries. Microsurgery 2008, 28(6):436-440.

[93] Pogrel MA, Maghen A: The use of autogenous vein grafts for inferior alveolar and lingual nerve reconstruction. Journal of oral and maxillofacial surgery : official journal of the American Association of Oral and Maxillofacial Surgeons 2001, 59(9): 985-988; discussion 988-993.

[94] Lavasani M, Gehrmann S, Gharaibeh B, Clark KA, Kaufmann RA, Peault B, Goitz RJ, Huard J: Venous graft-derived cells participate in peripheral nerve regeneration. PLoS One 2011, 6(9):e24801.

[95] Liard O, Segura S, Sagui E, Nau A, Pascual A, Cambon M, Darlix JL, Fusai T, Moyse E: Adult-brain-derived neural stem cells grafting into a vein bridge increases postlesional recovery and regeneration in a peripheral nerve of adult pig. Stem cells international 2012, 2012:128732.

[96] Norris RW, Glasby MA, Gattuso JM, Bowden RE: Peripheral nerve repair in humans using muscle autografts. A new technique. The Journal of bone and joint surgery British volume 1988, 70(4):530-533.

[97] Pereira JH, Palande DD, Narayanakumar TS, Subramanian AS, Gschmeissner S, Wilkinson M: Nerve repair by denatured muscle autografts promotes sustained sensory recovery in leprosy. The Journal of bone and joint surgery British volume 2008, 90(2): 220-224.

[98] Brunelli GA, Battiston B, Vigasio A, Brunelli G, Marocolo D: Bridging nerve defects with combined skeletal muscle and vein conduits. Microsurgery 1993, 14(4):247-251.

[99] Battiston B, Tos P, Cushway TR, Geuna S: Nerve repair by means of vein filled with muscle grafts I. Clinical results. Microsurgery 2000, 20(1):32-36.

[100] Stanec S, Stanec Z: Reconstruction of upper-extremity peripheral-nerve injuries with ePTFE conduits. Journal of reconstructive microsurgery 1998, 14(4):227-232.

[101] Rosson GD, Williams EH, Dellon AL: Motor nerve regeneration across a conduit. Microsurgery 2009, 29(2):107-114.

[102] Rinker B, Liau JY: A prospective randomized study comparing woven polyglycolic acid and autogenous vein conduits for reconstruction of digital nerve gaps. The Journal of hand surgery 2011, 36(5):775-781.

[103] Taras JS, Jacoby SM, Lincoski CJ: Reconstruction of digital nerves with collagen conduits. The Journal of hand surgery 2011, 36(9):1441-1446.

[104] Thomsen L, Bellemere P, Loubersac T, Gaisne E, Poirier P, Chaise F: Treatment by collagen conduit of painful post-traumatic neuromas of the sensitive digital nerve: a retrospective study of 10 cases. Chirurgie de la main 2010, 29(4):255-262.

[105] Moore AM, Kasukurthi R, Magill CK, Farhadi HF, Borschel GH, Mackinnon SE: Limitations of conduits in peripheral nerve repairs. Hand 2009, 4(2):180-186.

[106] Wangensteen KJ, Kalliainen LK: Collagen tube conduits in peripheral nerve repair: a retrospective analysis. Hand 2010, 5(3):273-277.

Chapter 3

Activity-Based Strategies in the Rehabilitation of Peripheral Nerve Injuries

Paulo A.S. Armada-da-Silva, C. Pereira, S. Amado,
A. Luís and A.C. Maurício

Additional information is available at the end of the chapter

http://dx.doi.org/10.5772/58437

1. Introduction

Peripheral nerve injuries are an important cause of permanent disability and have a strong negative impact on patients' quality of life. The neurological sequels of peripheral nerve injuries impair daily living and work activities and are often associated with serious complications, such as neuropathic pain [1]. The incidence of traumatic peripheral nerve injuries is higher in young male adults as a result of traffic, occupational and sport accidents [2], with the majority of injuries affecting the upper extremity [1], and a minor, though significant percentage of the nerve injuries damaging the sciatic nerve [3]. Although important advances were made in the surgical treatment, functional outcome following peripheral nerve injury is unsatisfying in most patients.

Experimental work with animal models has revealed several neurobiological mechanisms that are crucial for peripheral nerve regeneration and target reinnervation. This research highlights the role played by cellular and molecular mechanisms in signaling neuron injury and activating nerve's regenerative response. Different strategies have been used to enhance nerve regeneration, which use natural activities to stimulate the damaged nervous system to regenerate, with emphasis on treadmill exercise carried out in the immediate period following nerve damage and repair. In this Chapter we will review some of the work done recently regarding the utilization of activity-based strategies on axonal regeneration, reinnervation and functional recovery, as well as some of the mechanisms underlying these effects. Clinical evidence of the use of exercise and related treatment modalities in the rehabilitation of peripheral nerve damage will also be addressed. The importance of translating animal research to clinical settings and of developing new approaches for rehabilitation after nerve injury that comply with increasing knowledge about neurobiological mechanisms of peripheral nerve regeneration and functional recovery, and explore the plasticity of the central nervous system, will be mentioned.

2. Neuron early regenerative response

2.1. Injury signaling

Responses in the nerve at the site of injury begin almost immediately following axonal injury. Axotomy initiates a complex and coordinated set of injury signals that convey to neuron soma information regarding the axon injury [4]. Early signals arrive to the cell body in the form of vigorous electrical spiking activity that is generated at the lesion site and propagates up to the neuron soma. Membrane depolarization bursts are accompanied by the opening of voltage-gated and ligand-regulated Na^{+} and Ca^{2+} channels and by large transients of intracellular Ca^{2+} concentration increase, which activate several different Ca^{2+}-dependent kinases and raise cAMP levels. The formation of the growth cone, the initial event required for axonal elongation, relies on Ca^{2+} acting as intracellular second messenger and on signaling through the mitogen-activated protein kinase (MAPK) pathway, as well as on protein kinase A (PKA) activation [5]. In addition, Ca^{2+} entering the tip of the axon triggers Ca^{2+}-dependent proteases and enhances protein turnover and cytoskeletal dynamics, promoting growth cone advancement [6].

Axon injury disrupts retrograde axonal transport and deprives the neuron cell body of target-derived molecules, which are supposed to repress the neuronal intrinsic growth ability when neurons are firmly contacting their target organs. Little is known about these negative injury signals but the transforming growth factor beta (TGF-β)/SMAD2/SMAD3 pathway is a candidate. For instance, in primary sensory neurons, growth is repressed by the constitutively expressed inhibitor 5 of protein phophatase 1, which interacts with type 1 TGF-β receptor and triggers activity in the TGF-β/SMAD pathway, a mechanism that is down-regulated by injury [7].

Positive injury signals also make an important contribution to support the regenerative response of neurons after axotomy. At the site of the axonal lesion, several kinases, cytokines and downstream effectors are activated and transported back to the neuron soma [4]. The transport of phosphorylated MAPK from injured axons to the cell body leads to expression of regeneration-associated genes [8]. Likewise, activation of the mammalian-target of rapamycin (mTOR) pathway also promotes neuron growth in injured peripheral nerves, either by rapidly causing phosphorylation of the ribosomal S6 protein, a downstream effector of the mTOR pathway, or by regulating the expression of growth-associated protein(GAP)-43 [9]. Several neural growth factors and cytokines increase in concentration within peripheral nerves in response to injury. These factors include the glycosylated protein (gp)130 cytokine family members: leukemia inhibitory factor (LIF), interleukin-6 (IL-6) and ciliary neurotrophic factor (CNTF) [4], as well as neurotrophins, such as brain-derived neurotrophic factor (BDNF) and respective receptors [10]. These factors also play a crucial role in neuron survival and axonal regeneration [11].

2.2. Wallerian degeneration

Wallerian degeneration refers to the tightly regulated disruption of the distal axon in response to the injury. Myelin disintegration, detachment and proliferation of Schwann cells, macrophages activation, and recruitment of blood borne immune cells accompany Wallerian degeneration and create the conditions in the distal nerve to support axonal growth [12]. Schwann cells, macrophages, and other phagocytes recruited from the blood circulation and entering the nerve through the opened nerve-blood barrier, remove myelin debris and associated inhibitory signals, allowing regenerating axons to penetrate into the distal nerve [13].

In addition to proliferating and phagocytozing myelin debris, Schwann cells also secrete neurotrophic factors that promote axon growth thorough autocrine/paracrine mechanisms, along with cytokines and chemokines that regulate the inflammatory response [12]. For instance, the secretion of pro-inflammatory cytokines (e.g. TNF-α, IL-1, and IL-6) induces activation and secretion of phospholipases (e.g. phospholipase A2) that further increase the production of cytosolic and extracellular signaling molecules that contribute to clearance of myelin remnants in the distal nerve [13].

3. Major limitations for nerve regeneration

Experimental in vivo research highlights three key causes for poor reinnervation: 1) axonal loss of regenerative response [14], 2) inability of denervated distal nerve to support axonal growth [15], and 3) severe muscle atrophy [16, 17]. Growing axons progressively lose their regenerative ability if disconnected with targets. In the rat, the number of regenerating motoneurons declines to around one-third during approximately the first 4 months following axotomy [14]. In this case, however, reinnervated muscles are able to recover from atrophy and muscle strength is regained, as a result of intramuscular nerve sprouting and motor unit enlargement [14]. Chronic axotomized neurons can be stimulated to regenerate by the immunosuppressant FK506, thus reduced regenerative capacity can be overcome by proper stimuli [18].

Contrary to chronic axotomy, distal nerve stump denervation and severe muscle atrophy impede full recovery. Distal nerves that remain denervated for few months display diminished ability to support axonal growth and muscle reinnervation. As a consequence, muscles are only partially reinnervated as a result of decreased number of regenerating axons (see Figure 1). Only about 10 percent of motoneurons are capable of regenerating across a chronically denervated distal nerve [15]. In addition, similar numbers of motoneurons grow across the nerve pathway and reinnervate the muscles, showing the chronic denervated distal nerve as the main reason for severely impaired axonal regeneration and muscle reinnervation. This is likely the result of a decline in the number of Schwann cells, together with diminished ability of these cells to stimulate axon elongation [19]. In fact, Schwann cells reactivation by TGF-β elevates their capacity to support axon regeneration [20].

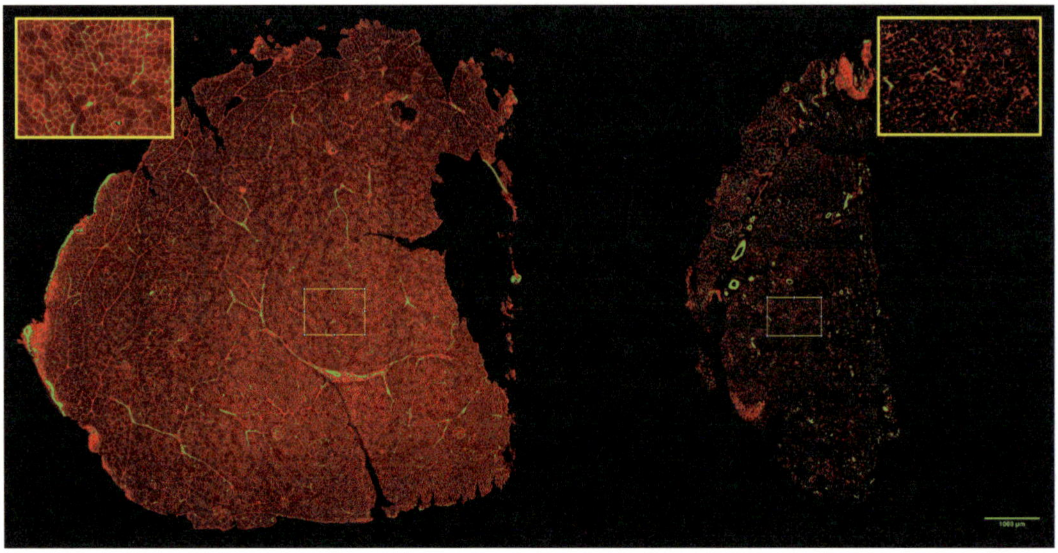

Figure 1. Montages of photomicrographs of transverse cryosections of the tibial anterior muscle from a control rat (left image) and from an animal 20 weeks following sciatic nerve transection and repair with a 10 mm-long nerve autograft (right image). Sections were immunostained for CD31 and fluorescein labeled Griffonia simplicifolia for visualization of blood vessels. Note the severe atrophy of the poorly reinnervated muscle and the extensive loss of capillaries. Scale bar: 1 mm.

Prolonged denervation leads to severe muscle atrophy, muscle fiber necrosis, fibrosis, and endplates disorganization [21, 22]. Extended denervated muscles when reinnervated by fresh axotomized nerves fail in restoring their tetanic force and muscle weight. The size of regenerated muscle units as well as muscle fibers' cross sectional area remain smaller when muscles stay denervated for extended periods of time [16]. Prolonged denervation initially causes atrophy of muscle fibers, but latter there is necrotic muscle changes that lessen the number of muscle fibers [21]. With time, denervated muscles also show impaired myogenesis, which further limits the ability to restore muscle mass even in the case muscles become reinnervated [23, 24] (Figure 2).

Delayed initiation of axonal growth plus misdirection of regenerating axons are additional reasons for poor functional recovery [25-27]. The injury site acts as a barrier that inhibits axonal regeneration and widens the interval of time that different neurons take to successfully initiate their growth (i.e. staggered regeneration). In the mean time, the distal nerve loses part of its ability to support axonal regeneration. Nerve injuries that preserve nerves' connective scaffold show better functional outcome [26]. In crush injuries the endoneurium remains intact along the entire distal nerve stump providing guidance for growing axons to reach their specific targets. Erratic guidance of regenerating axons increases if nerve injury disrupts the endoneurium. In this case, sprouts from regenerating axons may penetrate several different endoneurial pathways and terminate in targets that they formerly did not supply, as in the case of axons that regenerate within nerves leading to a totally different organ than the original (e.g. skin instead of muscle), or branches of a single motor axon ending up reinnervating muscles with antagonistic function.

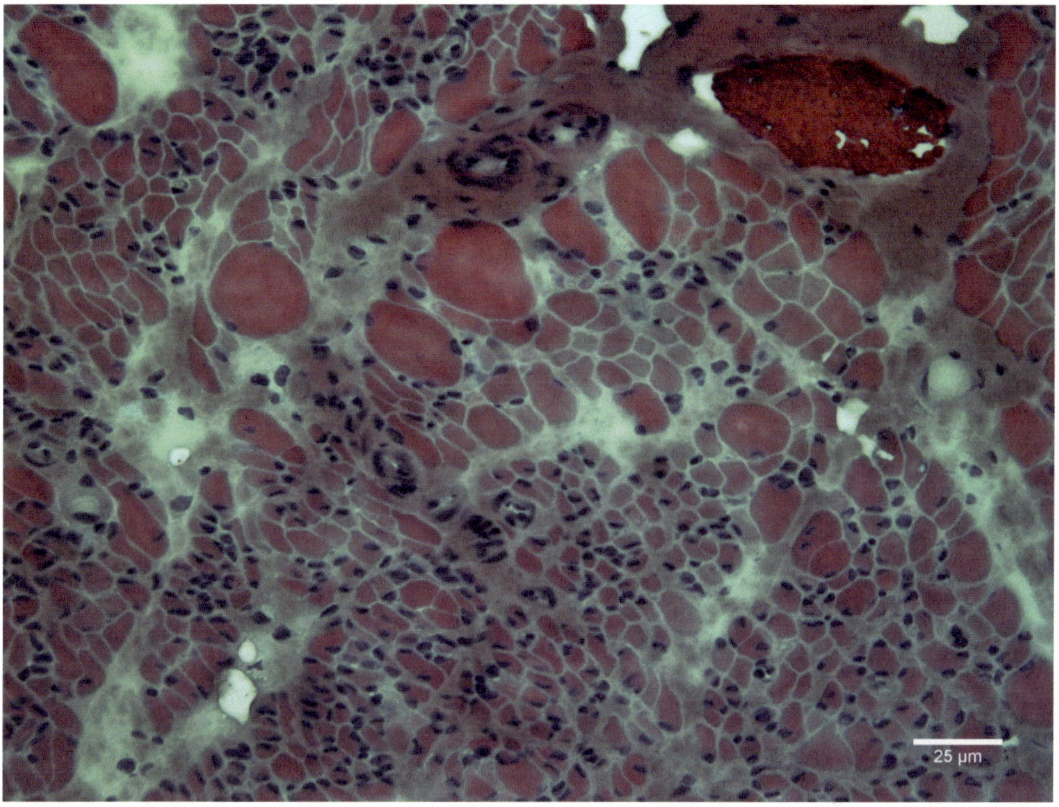

Figure 2. Photomicrograph of hematoxylin-eosin stained transverse cryosection of a poorly reinnervated tibial anterior muscle 20 weeks following sciatic nerve transection and repair with 10 mm-long nerve autograft. Visible a large number of highly atrophied muscle fibers with increased density of myonuclei, intermingled with small clusters of larger muscle fibers. Also visible, the large amount of connective tissue replacing the original muscle tissue. Scale bar: 25 microns.

4. Enhancement of axonal growth

4.1. Brief electrical stimulation

Stimulation of the neuronal activity early following axonal damage might strengthen the intrinsic injury signaling mechanisms and produce a more robust regenerative response. This hypothesis has been tested by using different methods to stimulate the activity of neurons soon following axotomy. Electrical stimulation of cut peripheral nerves applied immediately following axotomy is one of such methods. Brief one-hour, low-frequency electrical stimulation applied by the time of surgery accelerates motoneurons growth in the rat's femoral nerve and improves reinnervation specificity [28]. Importantly, repeating the electrical stimulation for two weeks postinjury does not result in additional positive effect on axonal regeneration [28]. Tetrodoxin abolishes the effect of electrical stimulation on axonal regeneration, which suggests that action potentials triggering and their propagation to the cell body are required to stimulate axonal regeneration by brief electrical stimulation. Thus, brief electrical stimulation applied immediately following axotomy probably strengthens intrinsic injury signaling, powerfully driving axonal regeneration.

Brief electrical stimulation acutely applied to the cut sciatic nerve also enhances motoneurons regeneration [29]. This effect is most noticed during the first two weeks postinjury, declining to some extent thereafter. However, in the cut and repaired sciatic nerve, electrical stimulation does not improve regeneration precision, therefore a significant proportion of regenerating motoneurons grow along the incorrect pathway, possibly affecting functional outcome negatively [29].

4.2. Treadmill exercise

Treadmill walking/running is the most commonly used form of activity-based experimental treatment within the context of peripheral nerve injury. An important question is whether treadmill exercise enhances axonal regeneration. Although initial studies addressing this question arrived to conflicting results (see ref. [30]), more recent evidence demonstrates that treadmill training accelerates axonal regeneration much in the same way that brief electrical stimulation does [31]. One hour of treadmill running conducted during the first two weeks (5 days/week) following sciatic nerve transection and direct repair in mice increases by four-fold the number of motoneurons that successfully regenerate during that period [32]. By the end of the fourth week of recovery, higher number of regenerated motoneurons could still be observed as a result of treadmill running performed two weeks earlier [32]. The enhancing effect of treadmill training on axonal regeneration can be achieved with different running protocols, ranging from continuous, mild-intensity, prolonged running (e.g. 1 hour/day, 10 m/min), to interval training made of bouts of high-intensity running, separated by periods of recovery (e.g. 4 x 2 min of running at 20 m/min and 5 min of recovery) [31, 33]. In addition to diminishing staggered regeneration, the more natural stimulation provided by treadmill exercise, relative to electrical stimulation, also prevents misrouting of the regenerating sciatic nerve motoneurons [32]. This effect might be crucial in terms of functional outcome considering that misrouting of regenerated neurons underlies pathological manifestations, such as dyskinesia [26]. Improved topographic organization of regenerated motoneurons in response to treadmill exercise is not clearly understood, but is likely related with better synchronized motoneuronal regeneration and higher competition for endoneurial pathways within the distal nerve stumps [32].

Resistance training also enhances regeneration of injured nerves [30]. This kind of exercise consists of performing sets of strong muscle contractions in order to increase muscle strength and augment muscle mass. Compared with resistance training and concurrent training (i.e. resistance and endurance training combined), as well as with sedentary controls, endurance training after sciatic nerve crush injury increases myelin sheath thickness of regenerated nerve fibers [34]. In addition, endurance training increases the percentage area of the regenerated nerve occupied by myelinated nerve fibers [34]. Resistance training also increases the diameter of the myelinated nerve fibers in crushed sciatic nerves, but only in the segment of the nerve proximal to the injury site [34].

Treadmill exercise and electrical stimulation can be combined, leading to a synergistic effect on promoting nerve regeneration and reinnervation [35]. In rats, and following sciatic nerve transection and direct co-optation, treadmill walking exercise (5 m/min) conducted during the initial four weeks of recovery increases the density and the number of myelinated nerve fibers in the tibial nerve [35]. The same effect is achieved by applying brief electrical stimulation at the time of the surgery. Contrariwise, chronic electrical stimulation fails in improving axonal regeneration. Combining treadmill exercise and brief electrical stimulation enhances the effect of the individual treatments. Low-intensity treadmill exercise in conjunction with acute electrical stimulation also produces faster and enhanced muscle reinnervation. A faster recovery of compound muscle action potential (i.e., M-wave) amplitude, as well as improved M-wave latency, are achieved if the acute electrical stimulation, which is delivered in a single one-hour session immediately after the nerve injury, is strengthened by treadmill exercise throughout the next four weeks. Interestingly, the effect of the combined acute electrical stimulation and treadmill exercise treatment is seen mostly in dorsiflexor muscles (e.g., tibialis anterior), which are supplied by the common peroneal branch of the sciatic nerve, compared to plantarflexors (e.g., plantaris muscle) that are innervated by the tibial nerve [35]. In addition, the positive effect of treadmill exercise on muscle reinnervation following sciatic nerve transection and repair is significant only past two months of injury and one month from the end of treadmill exercise [35, 36]. Once again, this suggests that treadmill exercise improves reinnervation, and possibly functional outcome as well, by acting upon early axonal regeneration, such as by diminishing staggered axonal regeneration and by raising the rate of axonal growth.

4.3. Passive mobilization

Passive mobilization is usually employed to maintain joint range of motion in paralyzed limbs as a result of peripheral nerve injury. In cases patients recover from total paralysis, as a result of successful reinnervation, passive mobilization is replaced by assisted mobilization in which the therapist aids patients moving their affected joint in the full range of motion. The aim of these treatments is to maintain joint function during the time damaged nerves regenerate and to increase muscle strength once reinnervation takes place.

However, passive mobilization might stimulate axonal regeneration as well. This has been demonstrated by experimental work showing improved end-plate structure, nerve sprouting, and end-plate reinnervation of extensor digitorum longus muscle as a result of passive mobilization undertaken during the immediate days post nerve injury [37]. Following facial nerve neurotmesis and direct cooptation in the rat, whisking function can be restored by stimulating passively the whisker pad for just few minutes daily [38]. Similar outcome is achievable in experimental injuries of the hypoglossal nerve [39]. Passive manual exercise does not alter the number of regenerated motoneurons or the topographic precision of reinnervation by the facial motoneurons [38]. However, passive activity of the denervated facial territories diminishes the number of Schwann cell bridges connecting end-plates of neighbor muscle fibers and the extent of muscle fibers poly-innervation [38]. Also, recovery of normal whisking function through passive manual treatment only occurs if trigeminal afferents are intact. In such cases, manual stimulation is associated with higher number of synaptic inputs onto facial motoneurons, suggesting that enhanced sensory stimulation achieved with manual stimulation is able to maintain appropriate levels of activity within the trigeminal-facial neural pathways [40].

The regeneration of sciatic motoneurons and muscle reinnervation of rat's hindleg muscles can also be promoted by passive cycling during the first weeks following sciatic nerve neurotmesis and end-to-end repair [36]. The effect of passive cycling on M-wave amplitude and latency, as well as in the magnitude of the electrically-elicited H-reflex, is comparable in magnitude to that of treadmill exercise [36]. These effects of passive mobilization are very relevant regarding translation to clinical practice, since patients are usually unable to undertake active physical exercise early following peripheral nerve injury.

5. Growth factors

The effect provided by activity-based strategies on axonal survival and regeneration may be linked with increased production and release of neurotrophins, growth factors, and hormones.

5.1. Neurotrophins

Neurotrophins are a family of extracellular signaling peptides that include the nerve growth factor (NGF), BDNF and neurotrophin(NT)-3 and NT-4/5. These neurotrophic factors bind to specific high-affinity tropomyosin-receptor kinase (Trk) receptors and to the low affinity p75 receptor. TrkA is the high affinity receptor for NGF, TrkB is the receptor for BDNF and NT-4/5, and TrkC is the receptor for NT-3 [41]. Trk and p75 receptors trigger different downstream intracellular pathways and different cell responses. Low affinity p75 receptors are usually up-regulated after injury and they seem to hamper axonal regeneration [11].

BDNF plays an important role in mediating the effects of physical exercise on synaptic plasticity in the brain [42]. Likewise, BDNF plays a crucial role in axonal regeneration based on several lines of evidence. Following injury, BDNF and their receptors TrkB and p75 are up-regulated in motoneurons and in denervated distal nerve stump [43, 44], although with diffrerences in response magnitude and timing between the two places. In motoneurons, BDNF mRNA expression is rapidly induced following axotomy, then returning to baseline after a few days. In intact nerves, BDNF is expressed at very low levels but in response to nerve transection the amount of BDNF mRNA increases steeply and with a magnitude that varies between different nerves [10]. In response to nerve injury, TrkB gene expression also increases in facial and sciatic motoneurons. The increase in TrkB mRNA levels begin in the immediate days post injury, reaches a three-fold peak increase by the end of the first week, and remains elevated throughout the next three to four weeks [10]. p75 mRNA also is rapidly up-regulated in axotomized motoneurons. The time course of p75 mRNA response is similar to that of TrkB mRNA but the magnitude is three to four-fold higher [10]. Despite the fast and robust increase in levels of BDNF and respective receptors, exogenous BDNF not always improves regeneration. At low doses, exogenous BDNF does not produce a clear effect on axonal regeneration of acutely injured and repaired peripheral nerves though it is able to promote regeneration of chronic axotomized neurons [45]. Also, large doses of exogenous BDNF impair axonal regeneration, probably by signaling through p75 receptors [45]. In fact, work on transgenic mice provides evidence that the TrkB receptor is necessary for adequate axonal regeneration, whereas signaling through p75 receptor has the opposite effect [11]. Brief electrical stimulation delivered to the proximal nerve stump immediately postinjury produces a fast and large BDNF and TrkB mRNAs response [46].

In the brain and in the spinal cord, BDNF expression is up-regulated by voluntary physical exercise with a magnitude that is in close relationship with the distance traveled [47]. The level of BDNF mRNA in the cell soma, in axons of spinal motoneurons, and in soleus muscle also increases after only a few bouts of exercise [48]. Although there is no direct evidence that physical exercise could increase BDNF levels in axotomized neurons or in the distal nerve stump following injury, the effect of treadmill exercise in promoting axonal regeneration requires BDNF expression by parental motoneurons [49]. In fact, treadmill training allows axons to grow into allografts harvested from Schwann cells BDNF-/-transgenic mice, which otherwise does not occur. However, in transgenic mice whose motoneurons do not express BDNF, axons fail to regenerate into grafts from Schwann cells BDNF-/-donors even when stimulated by treadmill exercise, thus suggesting that treadmill running up-regulates BDNF expression in regenerating motoneurons and that this is, at least in part, the mechanism by which treadmill running stimulates axonal regeneration [49]. BDNF and TrkB expression is also required for the role of passive manual stimulation on facial muscles'reinnervation. In fact, heterozygous deficient BDNF and TrkB mice, unlike their wild type counterparts, are unable to respond favorably to the manual passive treatment [50].

The role of the other neurotrophins, namely NGF, NT-3, NT-4/5, in promoting axonal regeneration is less well established, but simply based on changes in their expression following nerve injury they likely play a less important role compared to BDNF. Following axotomy, expression of NT-3 and NT-4/5, as well as that of TrkC, is rapidly down-regulated in parent motoneurons and in the distal nerve stump, whereas NGF levels increase in the distal nerve stump [10]. Despite being down-regulated, NT-3 treatment increases the number of motoneurons that successfully regenerate through nerve gaps and improves muscle reinnervation, specifically in fast contracting muscles [51, 52].

5.2. IGF-1

Insulin-like growth factor-1 (IGF-1) regulates many of skeletal muscle responses to physical exercise [53]. Muscle fibers increase the expression of IGF-1 in response to contractile activity and mechanical loading. Acting in autocrine/paracrine fashion, IGF-1 regulates muscle protein turnover and proliferation, and survival and differentiation of muscle-resident stem cells, namely satellite cells [53]. Serum levels of IGF-1 also increase as a result of different types of exercise, including resistance and endurance exercise [54], due to release from the muscle, liver and possibly other non-hepatic tissues, stimulated by growth hormone and by the exercise itself.

Numerous studies confirm the role played by IGF-1 on peripheral nervous system regeneration [55, 56]. Besides being released by muscle fibers, IGF-1 is also expressed in motoneurons and Schwann cells. In the nervous system, IGF-1 has a neuroprotective action and promotes the regeneration of peripheral nerves [57]. *In vitro*, IGF-1 promotes neurite outgrowth of cultured motoneurons, whereas *in vivo* it encourages terminal sprouting in reinnervating muscles [58].

Physical exercise protects against distinct types of brain injury. Such role of physical exercise is believed to be due to a higher quantity of IGF-1 entering the brain as a result of its elevated serum levels by virtue of exercise [59]. Moreover, ischemic brain injury is associated with large decreases in IGF-1 levels in the sciatic nerve, spinal cord, and brain cortex, probably resulting from inactivity [60]. Intramuscular administrations of recombinant IGF-1 increases its levels in the nervous system and in muscles and diminishes brain cortical cell apoptosis and motor dysfunction [60].

IGF-1 is also necessary for whisking function recovery following facial nerve transection and direct cooptation [61]. In heterozygous IGF-1-deficient mice, manual stimulation is ineffective in promoting functional recovery following facial nerve damage [61]. Although it is not possible to conclude from studies conducted in IGF-1 deficient mice that manual stimulation raises IGF-1 levels in the muscle or nerve, they demonstrate the supportive role played by IGF-1 in restoring function following peripheral nerve damage.

5.3. Testosterone

Testosterone and other androgens are important for normal muscle function, particularly regulating muscle anabolism [62]. Androgens also regulate axonal regeneration [63]. The effect of androgens in axonal growth varies with the specific nerve, with androgens playing a more important role in the regeneration of motoneurons in the facial nerve, relative to the sciatic nerve [63]. Androgens act on axonal regeneration through mechanism associated with the androgen receptor, as well as by modulation of stress cells' response, particularly the inhibition of heat shock proteins [63]. In addition, treatment with testosterone propionate leads to increased expression of BDNF and TrkB receptor by regenerating facial motoneurons [64].

The positive effect of treadmill running on axonal regeneration seems to be regulated also by sexual steroids. In mice, the effect of treadmill running on axonal regeneration varies according to gender and treadmill running protocol [65]. In males, but not in females or castrated males, continuous one-hour running each day for two weeks following common fibular nerve cut and repair significantly elevates testosterone serum levels, as well as accelerates axonal elongation [65]. Nonetheless, interval training running, comprised by bouts of intense running interspersed by periods of recovery, promotes axonal regeneration in female mice, although this training leaves serum testosterone baseline levels unchanged in both genders [65]. The use of an aromatase inhibitor, thus blocking the conversion of testosterone or of its precursors into estradiol, also improves axonal regeneration of common fibular motoneurons in female mice [65], thus suggesting that the failure of continuous exercise in stimulating axonal regeneration in females might be linked to the conversion of testosterone to estradiol or, in alternative, to a direct inhibitory effect of the latter on axonal regeneration.

Notwithstanding the subtleties of the effect of androgens on peripheral nerve regeneration, serum testosterone also increases acutely in response to physical exercise in human subjects [66], and this might aid nerve function.

6. Functional recovery

Although the role of physical exercise in enhancing axonal regeneration seems well established, its effect on functional recovery is less clear. In experimental models of peripheral nerve injury, several different tests are usually employed to evaluate functional recovery, including neurophysiological evaluation of motor reinnervation, muscle force testing, and behavioral tests [67]. In the case of sciatic nerve injury, behavioral test based on footprints, such as the sciatic functional index are commonly used [68]. These tests are non-invasive, relatively simple to perform and suitable for testing at several different and sequential time points.

Although behavioral tests offer relevant data about the recovery process, they gather only limited information regarding movement patterns changes and therefore cannot fully assess functional recovery. Thus, methods for evaluating movement production accurately and to assess performance of complex tasks requiring sensorimotor integration, such as gait, are necessary in peripheral nerve injury research [69].

6.1. Gait analysis

In the rat model, the study of limb kinematics during gait is a powerful means to evaluate functional recovery following peripheral nerve injury. Data of segmental and inter-joint coordination patterns can be combined with recordings of the electromyographical (EMG) activity of muscles [70] and ground reaction forces data [71, 72], providing detailed analysis of movements, including knowledge of joint powers and of the role played by muscles and other forces in producing the recorded movements.

We have used gait analysis in several occasions to assess the effect of different interventions following sciatic nerve injury in the rat, including the use of different tubulization procedures [73], application of biomaterials [74], and use of cellular systems [75, 76].

Figure 3 shows plots of ankle joint kinematics during the gait cycle, including both the stance and swing phases, prior to sciatic nerve transection and repair and at the end of 2 and 20 weeks of recovery in groups of adult male Sprague-Dawley walking across a walkway. Severe changes in ankle kinematics are easily noticed in animals 2 weeks after sciatic nerve transection and repair, which of course are expected due to the paralysis of the muscles crossing this joint. These changes are seen during both the stance and the swing phases of the gait cycle. During the stance phase, and in intact animals, the ankle joint first moves into dorsiflexion, reaching peak dorsiflexion near midstance, next performing plantarflexion until the end of this phase. Two weeks after sciatic nerve injury, ankle peak dorsiflexion angle is greatly increased and there is no plantarflexion during the second half of the stance phase, as a result of the paralysis of the ankle plantarflexor muscles (see also Figure 4). During the swing phase, changes in ankle kinematics are even more pronounced. During this phase of the gait cycle, the paw is lifted from the ground, moved forwardly and then placed again on the ground for the next stepping. Therefore, rats perform a very fast ankle joint action by which they first retract the limb (necessary for paw ground clearance) making ankle dorsiflexion and next extend the limb (to advance and place the paw on the ground again), now doing ankle plantarflexion. This requires fast contractions of the muscles actuating the ankle joint and fine coordination between ankle movement and those of the other limb joints. Acutely sciatic-injured animals are unable to produce such brisk ankle movements during the swing phase and the typical kinematics of the ankle joint is replaced by a short-range and slow plantarflexion that possibly occurs passively. After 20 weeks recovery, ankle kinematics is still deeply altered, despite a slight recovery of ankle plantarflexion near the stance-to-swing transition (Figures 3 and 4). Nevertheless, ankle kinematics during the swing phase remains severely disrupted even by the end of long term recovery (Figures 3 and 4).

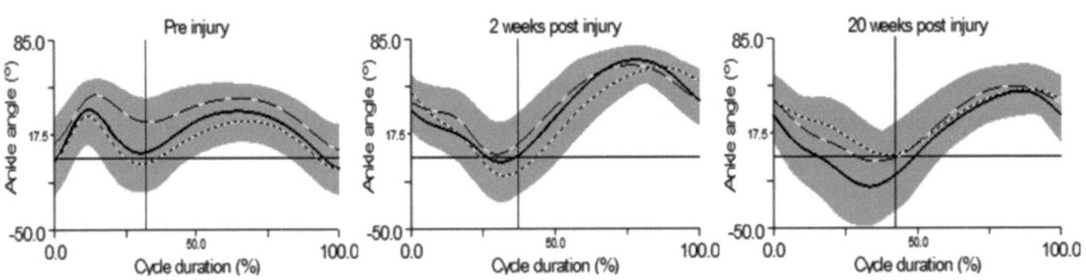

Figure 3. Plots of ankle joint kinematics during the gait cycle prior to sciatic nerve injury and at different times following sciatic nerve transection and repair. Curves represent groups of animals with their transected sciatic nerve treated with end-to-end repair, autograft or tubulization. Shadowed area depicts the standard deviation around the mean curves.

The poor recovery of ankle kinematics during gait following sciatic nerve transection and repair could be explained by limited motoneuron regeneration and lack of muscle reinnervation and strength recovery. However, this does not seem to be the case. In another occasion, we measured the torque produced by the ankle dorsiflexor and plantarflexor muscles in rats 16 weeks following sciatic nerve transection and repair and uninjured animals (unpublished observations; Figure 5). At the end of 16 weeks of recovery from sciatic nerve transection and repair, animals can produce relatively large dorsiflexor and plantarflexor torques. In addition, the torque-angle relationship for both reinnervated dorsiflexor and plantarflexor muscle groups remains largely unchanged.

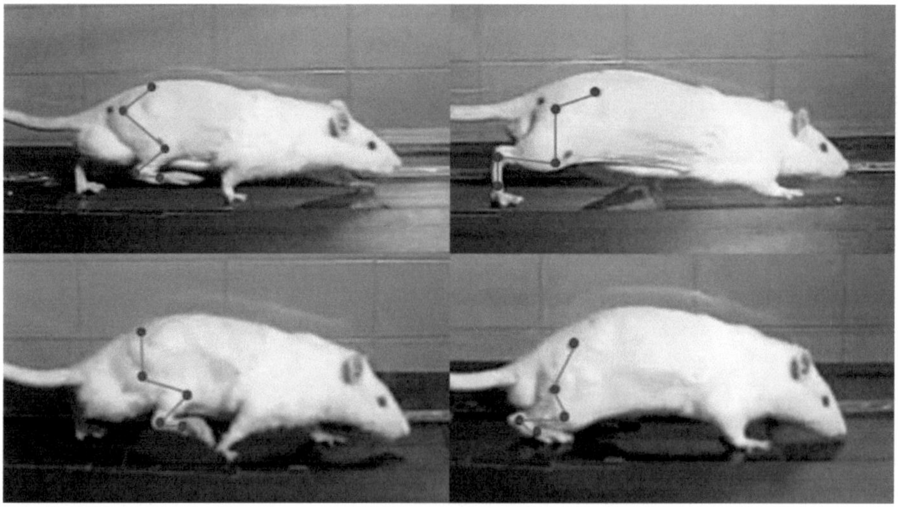

Figure 4. Photographs of an uninjured rat and a rat 16 weeks after sciatic nerve transection and end-to-end repair collected at approximate instants during the swing (images on the left side) and the stance (images on the right side) phases of the rat walking. Diagrams connecting the approximate centers of rotation of hip, knee, and ankle, as well as the fifth metatarsal head are superimposed. The diagrams are for illustration purposes only and are not supposed to accurately represent the joint angles. Impaired limb kinematics is visible in the sciatic-injured animal during both phases of the gait cycle.

The very limited recovery of normal gait pattern in rats following sciatic nerve transection despite significant recovery in muscle strength, demonstrates that axonal regeneration and muscle reinnervation, although necessary, are not sufficient for functional recovery. The concept of functional recovery is not always straightforward, particularly in the case of nervous system disorders. Following peripheral nerve injury, movement compensations emerge to respond to the disability and to maintain function, as for instance increased knee extension during walking to compensate the abnormal plantigrade gait [77] (Figure 4). Thereby, behavioral compensations are important for functional recovery, since they substitute some lost function, but they also mask disability and may be confounders when evaluating recovery after peripheral nerve injury, particularly when using more rudimentary tests [72]. Furthermore, the repeated use of movement compensations may turn them behaviorally fixed, thus eventually becoming an additional factor hampering further movement normalization [77].

6.2. Promoting spinal cord plasticity by activity-dependent strategies

Peripheral nerve injury causes permanent loss of muscle reflexes and triggers adaptive changes in the spinal cord and probably also in supraspinal sensorimotor centers that disrupt planning and ongoing regulation of movements [78]. Experimental studies with self-reinnervated single muscles illustrate well the role of changed muscle afferent feedback in regulating interjoint coordination and complex locomotor function [79]. Muscle self-reinnervation minimizes axonal misrouting and allows muscle reinnervation and muscle strength recovery. Likewise, hndlimb joint kinematics and patterns of muscle activity recover to normality despite the self-reinnerva-

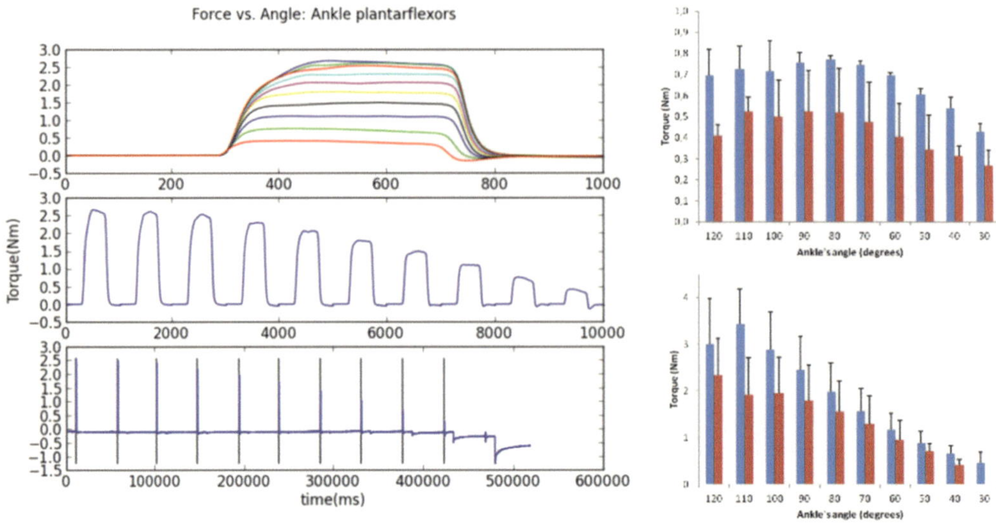

Figure 5. Left panel. Example of recordings of the torque produced during isometric tetanic contractions by the ankle plantarflexor muscles at different joint angles by an anesthetized rat 16 weeks after sciatic nerve transection and end-to-end repair. Ankle plantarflexor torque diminishes progressively with increased plantarflexion angle. In the lower graph, baseline down shifting close to the end of the recording indicates the passive torque generated by the soft tissue around the ankle joint near the end of the plantarflexion range of motion. Right panel. Mean values for ankle dorsiflexor (upper graph) and plantarflexor (lower graph) torque from 5 rats, 16 weeks following sciatic nerve transection and direct end-to-end repair and from equal number of uninjured control animals.

tion of ankle plantarflexors, but altered interjoint coordination appears if biomechanical constraints are imposed, such as walking up or down an incline or at a higher speed [79].

Proprioceptive deficits are the main explanation for changes in limb coordination after self-reinnervarion of ankle joint plantarflexors [78], and likely also in our sciatic nerve-injured animals. Reinnervated muscles are unresponsive to a stretch stimulus by virtue of unsuccessful reinnervation of muscle sensory organs by their specific sensory afferents, inability to rearrange central connections of sensory afferents to match changed target (e.g. Ib afferents changing their target from the Golgi end organ to muscle spindle receptors), and loss of monosynaptic sensory inputs onto motoneurons (i.e., synaptic stripping) [80, 81].

Spinal cord function, and in particular the recovery of central connections mediating muscle reflexes, may be ameliorated by up-conditioning of the H-reflex [82]. The strengthening of the H-reflex response using operant conditioning accelerates the recovery of the M-wave and H-reflex response in the soleus muscle in rats after sciatic nerve transection and repair [82]. Together with restoration of the electrical component of the muscle stretch reflex, H-reflex up-conditioning is associated anatomically with higher number of synaptic terminals established between primary sensory axons and motoneurons in the ventral horn [82].

These results are promising as they show the ability of activity-based interventions to shape spinal cord plasticity and revert, at least to some extent, nonadaptive secondary changes in spinal cord circuitry regulating motor output. A more normal kind of activity, in this case treadmill exercise also helps in restoring the H-reflex response in rats following sciatic nerve transection and repair, while also contributing to a pattern of muscle activation between antagonistic muscles of the ankle joint during gait better resembling that of uninjured animals [83].

7. Translational research and clinical studies

Although the positive effect of increased stimulation of motor and sensory pathways by natural activities, like treadmill exercise, on nerve regeneration, and possibly also on functional recovery, seems proven by the research conducted with animal models of peripheral nerve injury, similar evidence does not exist in the case of human patients. The lack of clear evidence demonstrating the efficacy of exercise therapy on functional outcomes is reported for conditions such as carpal tunnel syndrome [84], ulnar neuropathy [85], and Bell's palsy [86]. Nevertheless, in peripheral neuropathy, resistance exercise might increase muscle strength in affected muscles [87].

There are several physiotherapy modalities in the treatment of Bell's palsy which can be considered activity-based, like exercise therapy, biofeedback and mirror biofeedback, and relaxation. The results of several controlled randomized trials deny that any of such interventions brings a clear benefit to functional outcome [86, 88]. However, from a single preliminary study, there is evidence that active facial exercises are able to improve disability and diminish the prevalence of synkinesis in chronic facial palsy patients [86].

In carpal tunnel syndrome, exercise approaches are usually centered on nerve gliding and soft tissue mobilization. Other more holistic approaches, such as yoga, have also been attempted. In general, nerve gliding and stretching are considered effective in relieving symptoms of carpal tunnel syndrome by improving blood flow in the nerve, decreasing edema and pressure on the nerve, and mobilize the adherent medial nerve within the carpal tunnel [84]. Some studies report small size effect of exercise on measures of functional outcome in carpal tunnel syndrome, but taking into account studies' quality and the risk of bias, the overall evidence does not support a clear additional benefit of exercise or mobilization interventions on functional outcome in this condition [84]. Notwithstanding, carpal tunnel syndrome is a chronic condition, with periods of symptoms remission alternating with periods of symptoms exacerbation, therefore, with distinct physiopathology from that of peripheral nerve acute injury. However, preliminary results indicate that brief low-frequency electrical stimulation of the median nerve, applied in the perioperative period after median nerve releasing surgery, improves axonal regeneration and muscle reinnervation in carpal tunnel syndrome patients, but without clear improvement in terms of functional outcome [89].

8. Conclusion

Activity-based strategies improve nerve regeneration in rodent models of peripheral nerve injury. Natural stimulation of damaged motor and sensory pathways reinforces intrinsic neurobiological mechanisms of axonal growth, leading to faster and more complete target reinnervation. Central nervous system plasticity seems to be an important component of functional recovery of complex and adaptive sensorimotor behavior following peripheral nerve injury. Rat models may be useful to identify and develop novel activity-based strategies that stimulate axonal regeneration and central nervous system function, ultimately leading to improved functional outcome following peripheral nerve injury. Translation of such knowledge to clinical practice is desirable, although it must be carried out with caution and taking into consideration the clinical experience.

Acknowledgements

This research was supported by Project PTDC/DES/104036/2008, from the Portuguese Foundation for Science and Technology (Fundação Portuguesa para a Ciência e a Tecnologia), Portuguese Ministry of Science and Education (Ministério para a Ciência e Tecnologia) and by QREN I&DT Cluster in Development of Products for Regenerative Medicine and Cell Therapies – Projects Biomat& Cell QREN 2008/1372, co-financed by the European Community FEDER fund through ON2-O Novo Norte – North Portugal Regional Operational Program 2007-2013, by project "Hybrid Nanostructured Hydrogels: Bone regeneration using Multi-functional injectable Hydrogels-Rebone" –ENMED/0002/2010 from FCT, Ministério da Educação e da Ciência and Program Project Euronanomed, Ref: EraNet-EuroNanoMed JTC2010, and by the program COMPETE –Programa Operacional Factores de Competitividade, Project Pest-OE/AGR/UI0211/2011.

Author details

Paulo A.S. Armada-da-Silva[1,2], C. Pereira[1,2], S. Amado[2,3], A. Luís[4,5] and A.C. Maurício[4,5]

1 Department of Sports and Health, Faculdade de Motricidade Humana, Universidade de Lisboa, Cruz Quebrada, Portugal

2 Neuromechanics of Human Movement Group, Center for the Interdisciplinary Study of Human Performance (CIPER), Faculdade de Motricidade Humana, Universidade de Lisboa, Cruz Quebrada, Portugal

3 School of Health Sciences, Health Research Unit (UIS), Polytechnic Institute of Leiria, Portugal

4 Departamento de Clínicas Veterinárias, Instituto de Ciências Biomédicas de Abel Salazar (ICBAS), Universidade do Porto (UP), Porto, Portugal

5 Centro de Estudos de Ciência Animal (CECA), Instituto de Ciências e Tecnologias Agrarárias e Agro-Alimentares (ICETA), Porto, Portugal

References

[1] Ciaramitaro P, Mondelli M, Logullo F, Grimaldi S, Battiston B, Sard A, et al. Traumatic peripheral nerve injuries: epidemiological findings, neuropathic pain and quality of life in 158 patients. Journal of the Peripheral Nervous System. 2010;15(2): 120-127.

[2] Lad SP, Nathan JK, Schubert RD, Boakye M. Trends in median, ulnar, radial, and brachioplexus nerve injuries in the United States. Neurosurgery. 2010 May;66(5): 953-960.

[3] Maripuu A, Bjorkman A, Bjorkman-Burtscher I, Mannfolk P, Andersson G, Dahlin L. Reconstruction of sciatic nerve after traumatic injury in humans-factors influencing outcome as related to neurobiological knowledge from animal research. Journal of Brachial Plexus and Peripheral Nerve Injury. 2012;7(1):7; http://www.jbppni.com/content/7/1/7.

[4] Abe N, Cavalli V. Nerve injury signaling. Current Opinion in Neurobiology. 2008 Jun;18(3):276-283.

[5] Chierzi S, Ratto GM, Verma P, Fawcett JW. The ability of axons to regenerate their growth cones depends on axonal type and age, and is regulated by calcium, cAMP and ERK. The European Journal of Neuroscience. 2005 Apr;21(8):2051-2062.

[6] Verma P, Chierzi S, Codd AM, Campbell DS, Meyer RL, Holt CE, et al. Axonal Protein Synthesis and Degradation Are Necessary for Efficient Growth Cone Regeneration. The Journal of Neuroscience. 2005 January 12, 2005;25(2):331-342.

[7] Han Q-J, Gao N-N, Guo-QiangMa, Zhang Z-N, Yu W-H, Pan J, et al. IPP5 inhibits neurite growth in primary sensory neurons by maintaining TGF-β/Smad signaling. Journal of Cell Science. 2013 January 15, 2013;126(2):542-553.

[8] Agthong S, Kaewsema A, Tanomsridejchai N, Chentanez V. Activation of MAPK ERK in peripheral nerve after injury. BMC Neuroscience. 2006;7(1):45; http://www.biomedcentral.com/1471-2202/7/45.

[9] Abe N, Borson SH, Gambello MJ, Wang F, Cavalli V. Mammalian Target of Rapamycin (mTOR) Activation Increases Axonal Growth Capacity of Injured Peripheral Nerves. Journal of Biological Chemistry. 2010 September 3, 2010;285(36):28034-28043.

[10] Boyd JG, Gordon T. Neurotrophic factors and their receptors in axonal regeneration and functional recovery after peripheral nerve injury. Molecular Neurobiology. 2003 2003/06/01;27(3):277-323.

[11] Boyd JG, Gordon T. The neurotrophin receptors, trkB and p75, differentially regulate motor axonal regeneration. Journal of Neurobiology. 2001;49(4):314-325.

[12] Rotshenker S. Wallerian degeneration: the innate-immune response to traumatic nerve injury. Journal of Neuroinflammation. 2011;8:109; http://www.jneuroinflammation.com/content/8/1/109.

[13] Gaudet AD, Popovich PG, Ramer MS. Wallerian degeneration: gaining perspective on inflammatory events after peripheral nerve injury. Journal of Neuroinflammation. 2011;8:110; http://www.jneuroinflammation.com/content/8/1/110.

[14] Fu SY, Gordon T. Contributing factors to poor functional recovery after delayed nerve repair: prolonged axotomy. The Journal of Neuroscience. 1995 May;15(5 Pt 2): 3876-3885.

[15] Fu SY, Gordon T. Contributing factors to poor functional recovery after delayed nerve repair: prolonged denervation. The Journal of Neuroscience. 1995 May;15(5 Pt 2):3886-3895.

[16] Gordon T, Tyreman N, Raji MA. The basis for diminished functional recovery after delayed peripheral nerve repair. The Journal of Neuroscience. 2011 Apr 6;31(14): 5325-5334.

[17] Sulaiman W, Gordon T. Neurobiology of Peripheral Nerve Injury, Regeneration, and Functional Recovery: From Bench Top Research to Bedside Application. The Ochsner Journal. 2013 2013/03/01;13(1):100-108.

[18] Sulaiman OAR, Voda J, Gold BG, Gordon T. FK506 Increases Peripheral Nerve Regeneration after Chronic Axotomy but Not after Chronic Schwann Cell Denervation. Experimental Neurology. 2002;175(1):127-137.

[19] Sulaiman OAR, Midha R, Munro CA, Matsuyama T, Al-Majed A, Gordon T. Chronic Schwann Cell Denervation and the Presence of a Sensory Nerve Reduce Motor Axonal Regeneration. Experimental Neurology. 2002;176(2):342-354.

[20] Sulaiman OA, Gordon T. Role of chronic Schwann cell denervation in poor functional recovery after nerve injuries and experimental strategies to combat it. Neurosurgery. 2009 Oct;65(4 Suppl):A105-114.

[21] Borisov AB, Carlson BM. Cell death in denervated skeletal muscle is distinct from classical apoptosis. The Anatomical Record. 2000 Mar 1;258(3):305-318.

[22] Ijkema-Paassen J, Meek MF, Gramsbergen A. Reinnervation of muscles after transection of the sciatic nerve in adult rats. Muscle Nerve. 2002 Jun;25(6):891-897.

[23] Borisov AB, Dedkov EI, Carlson BM. Abortive myogenesis in denervated skeletal muscle: differentiative properties of satellite cells, their migration, and block of terminal differentiation. Anatomy and Embryology (Berl). 2005 Apr;209(4):269-279.

[24] Dedkov EI, Kostrominova TY, Borisov AB, Carlson BM. Reparative myogenesis in long-term denervated skeletal muscles of adult rats results in a reduction of the satellite cell population. The Anatomical Record. 2001 Jun 1;263(2):139-154.

[25] Brushart TM. Preferential reinnervation of motor nerves by regenerating motor axons. The Journal of Neuroscience. 1988 Mar;8(3):1026-1031.

[26] de Ruiter GC, Malessy MJ, Alaid AO, Spinner RJ, Engelstad JK, Sorenson EJ, et al. Misdirection of regenerating motor axons after nerve injury and repair in the rat sciatic nerve model. Experimental Neurology. 2008 Jun;211(2):339-350.

[27] Brushart TM, Hoffman PN, Royall RM, Murinson BB, Witzel C, Gordon T. Electrical stimulation promotes motoneuron regeneration without increasing its speed or conditioning the neuron. The Journal of Neuroscience. 2002 Aug 1;22(15):6631-6638.

[28] Al-Majed AA, Neumann CM, Brushart TM, Gordon T. Brief electrical stimulation promotes the speed and accuracy of motor axonal regeneration. The Journal of Neuroscience. 2000 Apr 1;20(7):2602-2608.

[29] English AW. Enhancing axon regeneration in peripheral nerves also increases functionally inappropriate reinnervation of targets. Journal of Comparative Neurology. 2005 Oct 3;490(4):427-441.

[30] van Meeteren NLU, Brakkee JH, Hamers FPT, Helders PJM, Gispen WH. Exercise training improves functional recovery and motor nerve conduction velocity after sciatic nerve crush lesion in the rat. Archives of Physical Medicine and Rehabilitation. 1997;78(1):70-77.

[31] English AW, Wilhelm JC, Sabatier MJ. Enhancing recovery from peripheral nerve injury using treadmill training. Annals of Anatomy-Anatomischer Anzeiger. 2011;193(4):354-361.

[32] English AW, Cucoranu D, Mulligan A, Sabatier M. Treadmill training enhances axon regeneration in injured mouse peripheral nerves without increased loss of topographic specificity. Journal of Comparative Neurology. 2009 Nov 10;517(2):245-255.

[33] Sabatier MJ, Redmon N, Schwartz G, English AW. Treadmill training promotes axon regeneration in injured peripheral nerves. Experimental Neurology. 2008 Jun;211(2): 489-493.

[34] Ilha J, Araujo RT, Malysz T, Hermel EE, Rigon P, Xavier LL, et al. Endurance and resistance exercise training programs elicit specific effects on sciatic nerve regeneration after experimental traumatic lesion in rats. Neurorehabilitation and Neural Repair. 2008 Jul-Aug;22(4):355-366.

[35] Asensio-Pinilla E, Udina E, Jaramillo J, Navarro X. Electrical stimulation combined with exercise increase axonal regeneration after peripheral nerve injury. Experimental Neurology. 2009 Sep;219(1):258-265.

[36] Udina E, Puigdemasa A, Navarro X. Passive and active exercise improve regeneration and muscle reinnervation after peripheral nerve injury in the rat. Muscle Nerve. 2011 Apr;43(4):500-509.

[37] Pachter BR, Eberstein A. Passive exercise and reinnervation of the rat denervated extensor digitorum longus muscle after nerve crush. American Journal of Physical Medicine and Rehabilitation. 1989 Aug;68(4):179-182.

[38] Angelov DN, Ceynowa M, Guntinas-Lichius O, Streppel M, Grosheva M, Kiryakova SI, et al. Mechanical stimulation of paralyzed vibrissal muscles following facial nerve injury in adult rat promotes full recovery of whisking. Neurobiology of Disease. 2007 Apr;26(1):229-242.

[39] Evgenieva E, Schweigert P, Guntinas-Lichius O, Pavlov S, Grosheva M, Angelova S, et al. Manual stimulation of the suprahyoid-sublingual region diminishes polyinnervation of the motor endplates and improves recovery of function after hypoglossal nerve injury in rats. Neurorehabilitation and Neural Repair. 2008 Nov-Dec;22(6): 754-768.

[40] Pavlov SP, Grosheva M, Streppel M, Guntinas-Lichius O, Irintchev A, Skouras E, et al. Manually-stimulated recovery of motor function after facial nerve injury requires intact sensory input. Experimental Neurology. 2008 May;211(1):292-300.

[41] Allodi I, Udina E, Navarro X. Specificity of peripheral nerve regeneration: interactions at the axon level. Progress in Neurobiology. 2012 Jul;98(1):16-37.

[42] Vaynman S, Ying Z, Gomez-Pinilla F. Interplay between brain-derived neurotrophic factor and signal transduction modulators in the regulation of the effects of exercise on synaptic-plasticity. Neuroscience. 2003;122(3):647-657.

[43] Funakoshi H, Frisén J, Barbany G, Timmusk T, Zachrisson O, Verge VM, et al. Differential expression of mRNAs for neurotrophins and their receptors after axotomy of the sciatic nerve. The Journal of Cell Biology. 1993 October 15, 1993;123(2):455-465.

[44] Kobayashi NR, Bedard AM, Hincke MT, Tetzlaff W. Increased Expression of BDNF and trkB mRNA in Rat Facial Motoneurons after Axotomy. European Journal of Neuroscience. 1996;8(5):1018-1029.

[45] Boyd JG, Gordon T. A dose-dependent facilitation and inhibition of peripheral nerve regeneration by brain-derived neurotrophic factor. European Journal of Neuroscience. 2002;15(4):613-626.

[46] Al-Majed AA, Brushart TM, Gordon T. Electrical stimulation accelerates and increases expression of BDNF and trkB mRNA in regenerating rat femoral motoneurons. European Journal of Neuroscience. 2000;12(12):4381-4390.

[47] Ying Z, Roy RR, Edgerton VR, Gomez-Pinilla F. Exercise restores levels of neurotrophins and synaptic plasticity following spinal cord injury. Experimental Neurology. 2005 Jun;193(2):411-419.

[48] Gomez-Pinilla F, Ying Z, Opazo P, Roy RR, Edgerton VR. Differential regulation by exercise of BDNF and NT-3 in rat spinal cord and skeletal muscle. The European Journal of Neuroscience. 2001 Mar;13(6):1078-1084.

[49] Wilhelm JC, Xu M, Cucoranu D, Chmielewski S, Holmes T, Lau K, et al. Cooperative Roles of BDNF Expression in Neurons and Schwann Cells Are Modulated by Exercise to Facilitate Nerve Regeneration. The Journal of Neuroscience. 2012 April 4, 2012;32(14):5002-5009.

[50] Sohnchen J, Grosheva M, Kiryakova S, Hubbers CU, Sinis N, Skouras E, et al. Recovery of whisking function after manual stimulation of denervated vibrissal muscles requires brain-derived neurotrophic factor and its receptor tyrosine kinase B. Neuroscience. 2010 Sep 29;170(1):372-380.

[51] Sterne GD, Brown RA, Green CJ, Terenghi G. Neurotrophin-3 Delivered Locally via Fibronectin Mats Enhances Peripheral Nerve Regeneration. European Journal of Neuroscience. 1997;9(7):1388-1396.

[52] Sterne GD, Coulton GR, Brown RA, Green CJ, Terenghi G. Neurotrophin-3–enhanced Nerve Regeneration Selectively Improves Recovery of Muscle Fibers Expressing Myosin Heavy Chains 2b. The Journal of Cell Biology. 1997 November 3, 1997;139(3):709-715.

[53] Goldspink G. Mechanical signals, IGF-I gene splicing, and muscle adaptation. Physiology (Bethesda). 2005 Aug;20:232-238.

[54] Hameed M, Lange KHW, Andersen JL, Schjerling P, Kjaer M, Harridge SDR, et al. The effect of recombinant human growth hormone and resistance training on IGF-I mRNA expression in the muscles of elderly men. The Journal of Physiology. 2004 February 15, 2004;555(1):231-240.

[55] Kim B, Leventhal PS, Saltiel AR, Feldman EL. Insulin-like Growth Factor-I-mediated Neurite Outgrowth in Vitro Requires Mitogen-activated Protein Kinase Activation. Journal of Biological Chemistry. 1997 August 22, 1997;272(34):21268-21273.

[56] Sullivan KA, Kim B, Feldman EL. Insulin-Like Growth Factors in the Peripheral Nervous System. Endocrinology. 2008 December 1, 2008;149(12):5963-5971.

[57] Kanje M, Skottner A, Sjo°berg J, Lundborg Gr. Insulin-like growth factor I (IGF-I) stimulates regeneration of the rat sciatic nerve. Brain Research. 1989;486(2):396-398.

[58] Caroni P, Schneider C, Kiefer MC, Zapf J. Role of muscle insulin-like growth factors in nerve sprouting: suppression of terminal sprouting in paralyzed muscle by IGF-binding protein 4. The Journal of Cell Biology. 1994 May 15, 1994;125(4):893-902.

[59] Carro E, Trejo JL, Busiguina S, Torres-Aleman I. Circulating Insulin-Like Growth Factor I Mediates the Protective Effects of Physical Exercise against Brain Insults of Different Etiology and Anatomy. The Journal of Neuroscience. 2001 August 1, 2001;21(15):5678-5684.

[60] Chang H-C, Yang Y-R, Wang PS, Kuo C-H, Wang R-Y. The Neuroprotective Effects of Intramuscular Insulin-Like Growth Factor-I Treatment in Brain Ischemic Rats. PLoS One. 2013;8(5):e64015; http://dx.doi.org/10.1371%2Fjournal.pone.0064015.

[61] Kiryakova S, Sohnchen J, Grosheva M, Schuetz U, Marinova T, Dzhupanova R, et al. Recovery of whisking function promoted by manual stimulation of the vibrissal muscles after facial nerve injury requires insulin-like growth factor 1 (IGF-1). Experimental Neurology. 2010 Apr;222(2):226-234.

[62] Griggs RC, Kingston W, Jozefowicz RF, Herr BE, Forbes G, Halliday D. Effect of testosterone on muscle mass and muscle protein synthesis. Journal of Applied Physiology. 1989 Jan;66(1):498-503.

[63] Jones KJ, Brown TJ, Damaser M. Neuroprotective effects of gonadal steroids on regenerating peripheral motoneurons. Brain Research Reviews. 2001;37(1–3):372-382.

[64] Sharma N, Marzo SJ, Jones KJ, Foecking EM. Electrical stimulation and testosterone differentially enhance expression of regeneration-associated genes. Experimental Neurology. 2010;223(1):183-191.

[65] Wood K, Wilhelm JC, Sabatier MJ, Liu K, Gu J, English AW. Sex differences in the effectiveness of treadmill training in enhancing axon regeneration in injured peripheral nerves. Developmental Neurobiology. 2012;72(5):688-698.

[66] Ahtiainen JP, Pakarinen A, Alen M, Kraemer WJ, Hakkinen K. Muscle hypertrophy, hormonal adaptations and strength development during strength training in strength-trained and untrained men. European Journal of Applied Physiology. 2003 Aug;89(6):555-563.

[67] Wood MD, Kemp SW, Weber C, Borschel GH, Gordon T. Outcome measures of peripheral nerve regeneration. Annals of Anatomy. 2011 Jul;193(4):321-333.

[68] Varejao AS, Cabrita AM, Meek MF, Bulas-Cruz J, Melo-Pinto P, Raimondo S, et al. Functional and morphological assessment of a standardized rat sciatic nerve crush injury with a non-serrated clamp. Journal of Neurotrauma. 2004 Nov;21(11): 1652-1670.

[69] Joao F, Amado S, Veloso A, Armada-da-Silva P, Mauricio AC. Anatomical reference frame versus planar analysis: implications for the kinematics of the rat hindlimb during locomotion. Reviews in the Neurosciences. 2010;21(6):469-485.

[70] Sabatier MJ, To BN, Nicolini J, English AW. Effect of Axon Misdirection on Recovery of Electromyographic Activity and Kinematics after Peripheral Nerve Injury. Cells Tissues Organs. 2011 Mar 17;193(5):298-309.

[71] Howard CS, Blakeney DC, Medige J, Moy OJ, Peimer CA. Functional assessment in the rat by ground reaction forces. Journal of Biomechanics. 2000 Jun;33(6):751-757.

[72] Bennett SW, Lanovaz JL, Muir GD. The biomechanics of locomotor compensation after peripheral nerve lesion in the rat. Behavioral and Brain Research. 2012 Apr 15;229(2):391-400.

[73] Luis AL, Rodrigues JM, Geuna S, Amado S, Shirosaki Y, Lee JM, et al. Use of PLGA 90:10 Scaffolds Enriched with In Vitro-Differentiated Neural Cells for Repairing Rat Sciatic Nerve Defects. Tissue Engineering Part A. 2008 Jun;14(6):979-993.

[74] Amado S, Rodrigues JM, Luis AL, Armada-da-Silva PA, Vieira M, Gartner A, et al. Effects of collagen membranes enriched with in vitro-differentiated N1E-115 cells on rat sciatic nerve regeneration after end-to-end repair. Journal of Neuroengineering and Rehabilitations. 2010;7:7; http://www.jneuroengrehab.com/content/7/1/7.

[75] Amado S, Simoes MJ, Armada da Silva PA, Luis AL, Shirosaki Y, Lopes MA, et al. Use of hybrid chitosan membranes and N1E-115 cells for promoting nerve regeneration in an axonotmesis rat model. Biomaterials. 2008 Nov;29(33):4409-4419.

[76] Gartner A, Pereira T, Armada-da-Silva PA, Amorim I, Gomes R, Ribeiro J, et al. Use of poly(DL-lactide-epsilon-caprolactone) membranes and mesenchymal stem cells from the Wharton's jelly of the umbilical cord for promoting nerve regeneration in axonotmesis: in vitro and in vivo analysis. Differentiation. 2012 Dec;84(5):355-365.

[77] Amado S, Armada-da-Silva PA, Joao F, Mauricio AC, Luis AL, Simoes MJ, et al. The sensitivity of two-dimensional hindlimb joint kinematics analysis in assessing functional recovery in rats after sciatic nerve crush. Behavioural Brain Research. 2011 Dec 1;225(2):562-573.

[78] Alvarez FJ, Bullinger KL, Titus HE, Nardelli P, Cope TC. Permanent reorganization of Ia afferent synapses on motoneurons after peripheral nerve injuries. Annals of the New York Academy of Sciences. 2010;1198(1):231-241.

[79] Maas H, Prilutsky BI, Nichols TR, Gregor RJ. The effects of self-reinnervation of cat medial and lateral gastrocnemius muscles on hindlimb kinematics in slope walking. Experimental Brain Research. 2007 Aug;181(2):377-393.

[80] Alvarez FJ, Titus-Mitchell HE, Bullinger KL, Kraszpulski M, Nardelli P, Cope TC. Permanent central synaptic disconnection of proprioceptors after nerve injury and regeneration. I. Loss of VGLUT1/IA synapses on motoneurons. Journal of Neurophysiology. 2011 November 1, 2011;106(5):2450-2470.

[81] Bullinger KL, Nardelli P, Pinter MJ, Alvarez FJ, Cope TC. Permanent central synaptic disconnection of proprioceptors after nerve injury and regeneration. II. Loss of functional connectivity with motoneurons. Journal of Neurophysiology. 2011 November 1, 2011;106(5):2471-2485.

[82] Chen Y, Wang Y, Chen L, Sun C, English AW, Wolpaw JR, et al. H-reflex up-conditioning encourages recovery of EMG activity and H-reflexes after sciatic nerve transection and repair in rats. The Journal of Neuroscience. 2010 Dec 1;30(48):16128-16136.

[83] Boeltz T, Ireland M, Mathis K, Nicolini J, Poplavski K, Rose SJ, et al. Effects of treadmill training on functional recovery following peripheral nerve injury in rats. Journal of Neurophysiology. 2013 June 1, 2013;109(11):2645-2657.

[84] Page MJ, O'Connor D, Pitt V, Massy-Westropp N. Exercise and mobilisation interventions for carpal tunnel syndrome. Cochrane Database of Systematic Reviews. 2012;6:CD009899.

[85] Caliandro P, La Torre G, Padua R, Giannini F, Padua L. Treatment for ulnar neuropathy at the elbow. Cochrane Database of Systematic Reviews. 2012;7:CD006839.

[86] Teixeira LJ, Valbuza JS, Prado GF. Physical therapy for Bell's palsy (idiopathic facial paralysis). Cochrane Database Syst Rev. 2011(12):CD006283.

[87] White Claire M, Pritchard J, Turner-Stokes L. Exercise for people with peripheral neuropathy. Cochrane Database of Systematic Reviews [serial on the Internet]. 2004;(4).

[88] Beurskens CHG, Burgers-Bots IAL, Kroon DW, Oostendorp RAB. Literature Review of Evidence Based Physiotherapy in Patients with Facial Nerve Paresis. Journal of the Japanese Physical Therapy Association. 2004;7(1):35-39.

[89] Gordon T, Amarjani N, Edwards DC, Chan KM. Brief post-surgical electrical stimulation accelerates axon regeneration and muscle reinnervation without affecting the functional measures in carpal tunnel syndrome patients. Experimental Neurology. 2009 Oct 1.

Chapter 4

Surgical Management of Obstetric Brachial Plexus Palsy Secondary Deformities

Rahul K. Nath and Chandra Somasundaram

Additional information is available at the end of the chapter

http://dx.doi.org/10.5772/58322

1. Introduction

The objectives of this book chapter are to describe the diagnostics, clinical assesment and surgical management of obstetric brachial plexus injury (OBPI) or brachial plexus birth palsy (BPBP).

OBPI occurs during delivery process. The incidence of OBPI has been reported to vary between 0.38 and 5.8 for every 1000 live births [1-5]. The occurrence has increased despite the advances in obstetrics, and medical technology [6, 7]. Reported risk factors for OBPI include shoulder dystocia, macrosomia (defined as birth weight greater than 4500 g) [8-11]), instrument-assisted delivery, and downward traction of the fetal head [3, 12, 13]. Shoulder dystocia is the most prevalent risk factor in our patients [14]; almost all the children in our study had documented shoulder dystocia [14]. Shoulder dystocia is, therefore, closely associated with the most severe cases of permanent obstetric brachial plexus injuries [8, 9, 11]. However, permanent injury is not exclusive to large infants; 80% of the OBPI patients in our published study were not macrosomic and 43% (104/241) weighed less than 4000 g at birth [14]. OBPI that occur during breech deliveries may have a different mechanism of onset, and are more likely to be bilateral. Avulsions of the upper roots are more likely during breech than during vertex delivery [15, 16]. OBPI may also occur, although very rarely, during cesarean sections [17].

The most commonly affected roots are C5–6 (Erb's palsy), because of their more superficial location in the neck, are more vulnerable to injury. Less frequently, the entire plexus (C5–T1) may be affected [18, 19]. The injury can be simple stretch or rupture or avulsion. Most of these injuries are transient; patients recover functions spontaneously within the 3 months of life. However, a significant proportion of these children tend to retain persistent limb deficits, never recover full function and develop permanent injuries [1, 20, 21]. Unlike adults, children may have complications from even the most simple nerve injury due to the growth issues that are present. The mildest and most common OBPI is neurapraxia; the most severe is avulsion [20, 22]. Both types of injury have the potential to result in permanent disability.

2. Diagnostics

Diagnostic tools used to identify which lesions are permanent in OBPI include computed tomography (CT), magnetic resonance imaging (MRI), myelogram, and electromyography (EMG) as well as nerve conduction velocity (NCV) studies [23-27]. Distinguishing preganglionic (avulsion) from postganglionic (rupture) lesions is critical, and can be difficult at initial presentation based on clinical examination alone in these infants [26, 28]. Our experience with MRI for pre-operative assessment of the spinal roots has been unfavorable. EMG testing is the procedure of choice for preoperative evaluation of nerve-muscle integrity.

2.1. Radiological evaluation

In order to assess bony deformities of the shoulder joint, CT or MRI images of the patients are studied before and after triangle tilt surgery. Posterior humeral head subluxation, glenoid version, and SHEAR deformity are measured from the radiographs (CT/MRI scans). Glenoid version (normal value=0) is measured as described by Friedman et al. [29] using axial CT/MRI images (Figure 1). A scapular line connecting the mid-glenoid to the medial spine of the scapula is constructed using Universal Desktop Ruler (AVPSoft.com, Voronezh, Russia). The angle formed between the scapular line and a line drawn tangential to the glenoid surface interacting closely with the humeral head is calculated and 90º is subtracted from it to measure the glenoscapular angle. Posterior subluxation of the humeral head (Figure 1) is expressed as percentage of humeral head anterior to the glenoid (PHHA, normal value=50), and calculated from the ratio of the distance between the scapular line to the anterior aspect of humeral head and the greatest diameter of the head, multiplied by 100.

The scapular deformity, also referred as SHEAR deformity is measured from the 3D reconstructions of the CT images (Figure 2). The area of the scapula visible above the clavicle is measured and divided with the total area of the scapula for both affected and normal sides. The ratio of the affected side is subtracted from that of the normal side and multiplied with 100 to obtain SHEAR deformity (normal value=0).

2.2. Clinical assessment

Shoulder function is assessed through the modified Mallet scale through video recordings of patients performing the following movements: shoulder abduction, external rotation, hands to mouth, hands to neck, hands to spine, and supination. For each functional Mallet parameter, patients are scored on a scale of 1–5 with 5 as normal function, and 1 denoting lack of any movement (Figure 3). Despite continuing improvements in diagnostic technology, at this time, the final diagnoses must be made during surgery in complex or unclear cases.

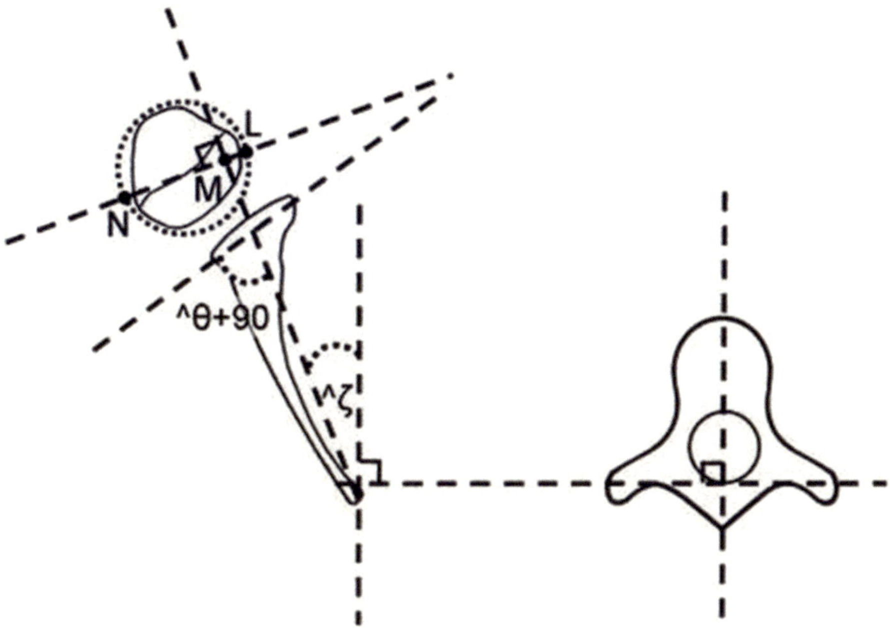

Figure 1. Schematic drawing showing the method of calculating glenoscapular angle (glenoid version θ), posterior subluxation of the humeral head and spinoscapular angle (ζ) [40].

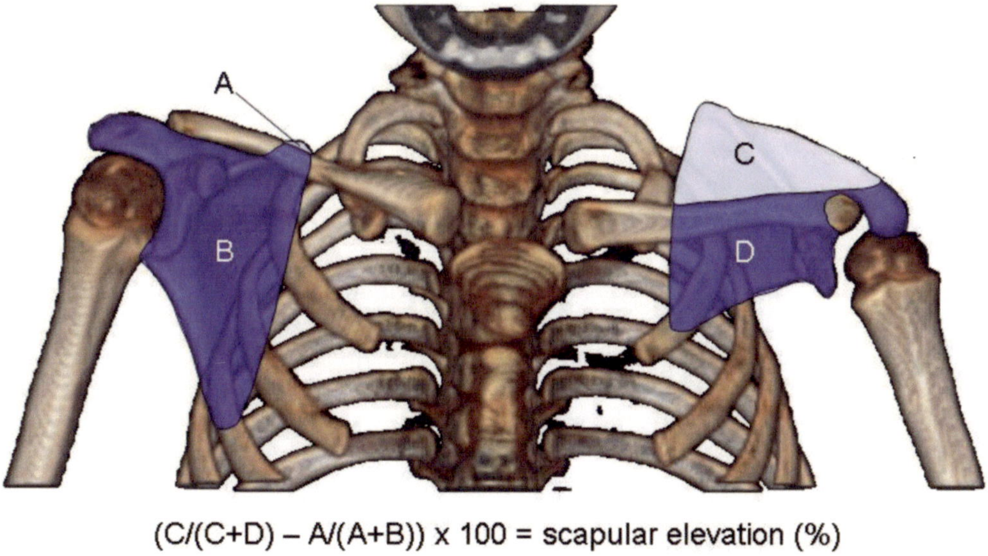

Figure 2. Measuring scapular elevation to quantitate the extent of the SHEAR deformity. Shown here is the CT for a patient with 37% scapular elevation [42].

Figure 3. Modified Mallet scale evaluation of function and arm appearance. In addition to assessing the classical shoulder functions of the classical Modified Mallet system, supination and the resting position are evaluated. [19]

3. Non-surgical interventions

Non-Surgical interventions are physical and occupational therapies, electrostimulation, neuromotor therapy, BTX-A injections and splinting. Among the most promising of these methods is BTX-A treatment. This has been shown to treat biceps/triceps co-contraction [30-32], and shown to improve biceps movement and strength [33]. BTX-A treatment for OBPI has not been effective over long-term, in our experience.

4. Primary surgical solutions

There are three primary surgical solutions: neurolysis, neurotization (nerve transfer) and nerve grafting. These may be performed alone or in combination with each other. The choice of which peripheral nerve surgery technique is appropriate is based on which method will maximize and encourage the natural regenerative process of the nerve. We use nerve transfer and grafting, depending on the intra-operative findings and with our surgical experiences. Injuries of C5-C6 do not always warrant surgery if injury to C7 is minor. In patients where C8 and T1 are involved in addition to C5-C7, hand function is also affected, therefore nerve repair for the upper and middle trunks is more likely to be required.

5. Surgical management

The decision of whether to surgically repair the nerves, however, does typically need to be made in the first 6 months of life. Delay beyond this age in these patients lead to long-term morbidity by causing muscle imbalances and weakness around the shoulder (the deltoid and external shoulder rotators) [34-37], and bony deformities at the shoulder joint (glenohumeral dysplasia and joint incongruity) [20, 38, 39]. These anatomical changes subsequently severely impair the bone growth and development [20]. The major bony deformity that develops is termed as the SHEAR (scapular hypoplasia, elevation, and rotation) deformity, which is caused by the elevation and extrusion of the affected scapula beyond the clavicle [40]. The abnormal anterior rotation of the clavicle together with the protracted scapula causes the acromioclavicular plane to tilt forward and thereby lead to the impingement of the acromion upon the humeral head [41, 42]. Significant secondary deformities that follow include medial rotation contracture (MRC) and elbow flexion.

Early surgical interventions have been shown to improve the limb functions in this group of patients [43, 44]. Management of secondary deformities in OBPI has typically been through the performance of various operative procedures including tendon transfers, muscle releases, axillary nerve decompression, humeral osteotomy, biceps tendon lengthening, glenohumeral capsulorrhaphy and anterior capsule release [45-51].

5.1. Z-lengthening

Biceps tendon lengthening/ the Z-lengthening is an option in C5-C7 (asymmetric) nerve injury, where the biceps recovers faster, thereby overpowering the triceps. The added length that is achieved allows straightening of the elbow and provides additional length to the arm [19].

5.2. Ilizarov bone lengthening

In severe OBPI patients, who is left with severe bony rotational and shortening deformities that are functionally limiting, the use of Ilizarov bone distraction technique is appropriate. This technique is used for rotation and lengthening of the humerus as well as the forearm. Functional gains are significant as the hand is placed into a more useful position [19].

5.3. Posterior glenohumeral capsulorrhaphy

Posterior glenohumeral capsulorrhaphy tightens the posterior capsule surrounding the humeral head and repositions it anteriorly. This procedure does not address the SHEAR deformity [40] and its central influence in the pathophysiology of the medial rotation contracture. In our experience, on its own, posterior capsulorrhaphy is often not sufficient to address the glenohumeral subluxation, as is predictable when taking the SHEAR into consideration. In our experience, successful restoration of position and function in failed humeral osteotomy patients has followed from surgically addressing the SHEAR deformity. It may be inferred that the SHEAR correction, the Triangle Tilt surgery is a more specific operation because it addresses the root cause of the medial rotation.

6. Muscle and bone deformities and their management

We have been less aggressive to nerve reconstruction and paying more attention to the secondary and tertiary consequences of the initial nerve injury, based on developing and quite compelling literature and on our own experience with several thousand patients [52]. We described our preferred management for muscle injury as a result of OBPI, with supporting clinical and literature evidence. The traditional muscle release operations do not adequately address the pathophysiology of the shoulder in OBPI patients in our experience. Therefore, the surgeon and the lead author (RKN) [53-55] has modified the previously described soft tissue release operation [56], by coupling neurolysis and decompression of the axillary nerve with an untethering release of soft tissue contractures (modified Quad, figure 4) [53-55]. We have demonstrated that modified Quad [53-55] lead to better shoulder abduction and flexion through releasing the existing contractures.

6.1. Modified Quad (figure 4)

1. Transfer of the latissimus dorsi muscle to give external rotation and abduction.
2. Transfer of the teres major muscle to stabilise the scapula.
3. Release of the subscapularis, pectoralis major and minor contractures.
4. Decompression and neurolysis of the axillary nerve [53, 54].

However, these procedures may not address the glenohumeral dysplasia and joint incongruity. Restoration of glenohumeral congruity is therefore a primary objective in treating OBPI, which then allows for maximum functional range of motion and improved limb growth. A bony surgical procedure, Triangle Tilt (figure 5) [42, 44, 57-66] was therefore developed by the lead author and surgeon (RKN).

6.2. The triangle tilt surgery consists of

1. osteotomy of the clavicle at the junction of the middle and distal thirds,
2. osteotomy of the acromion process at its junction with the spine of the scapula,
3. ostectomy of the superomedial angle of the scapula to reduce scapular winging,

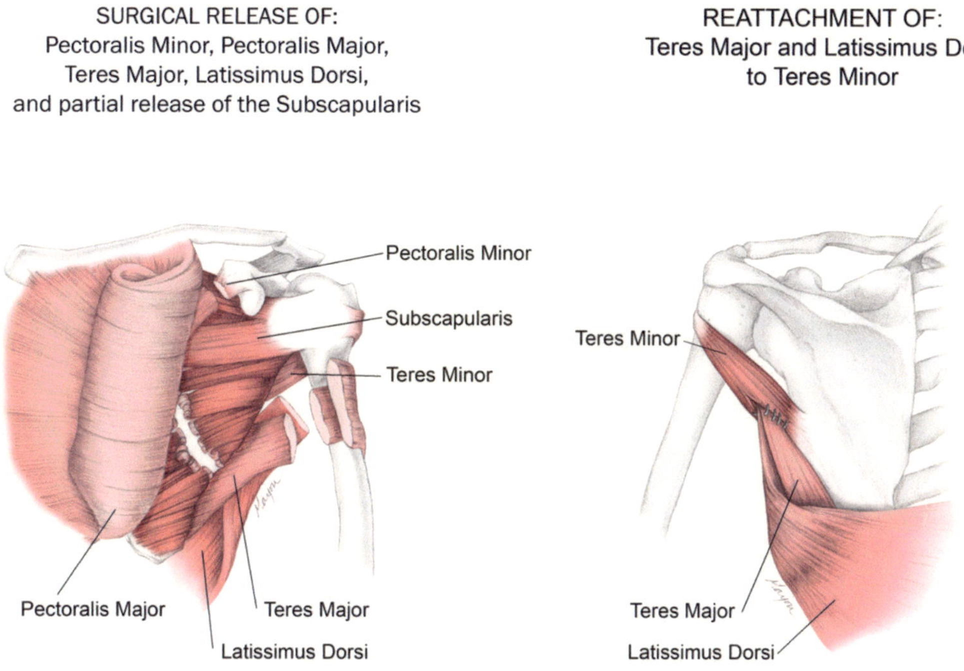

Figure 4. The Mod Quad Procedure improves shoulder abduction and flexion. Left, release of major internal rotator muscles: subscapularis (not shown), teres major, latissimus dorsi, pectoralis major and minor. Right, teres major and latissimus dorsi are transferred to the teres minor, increasing external rotation, abduction and scapular stability. Neurolysis and decompression of the axillary nerve further increase range of motion [19].

4. splinting of the extremity in adduction, external rotation and forearm supination.

This triangle tilt detaches the distal acromio-clavicular triangle-humeral head complex from the abnormally positioned scapula, and tilts the acromio-clavicular plane back to neutral position. This relieves the impingement of the acromio-clavicular triangle on the humeral head. and allows the head to be repositioned passively into a neutral position within the glenoid fossa (Figure 5), resulting in improved gleno-humeral joint congruency [42, 44, 57-66].

We have demonstrated the short (1 year), and long-term (2 years), and extended long term (5 years) benefits of triangle tilt surgery in OBPI patients (age between 0.9 and 17 year old) by examination of their radiological reports as well as the modified Mallet functional scale [42, 44, 57-66]. In addition, triangle tilt surgery is a salvage procedure in failed humeral osteotomy patients [58, 65]. Minor elements of the procedure include bone grafting of the acromion process and clavicular osteotomy sites and semi-rigid fixation of the clavicular osteotomy segments to prevent nonunion [19].

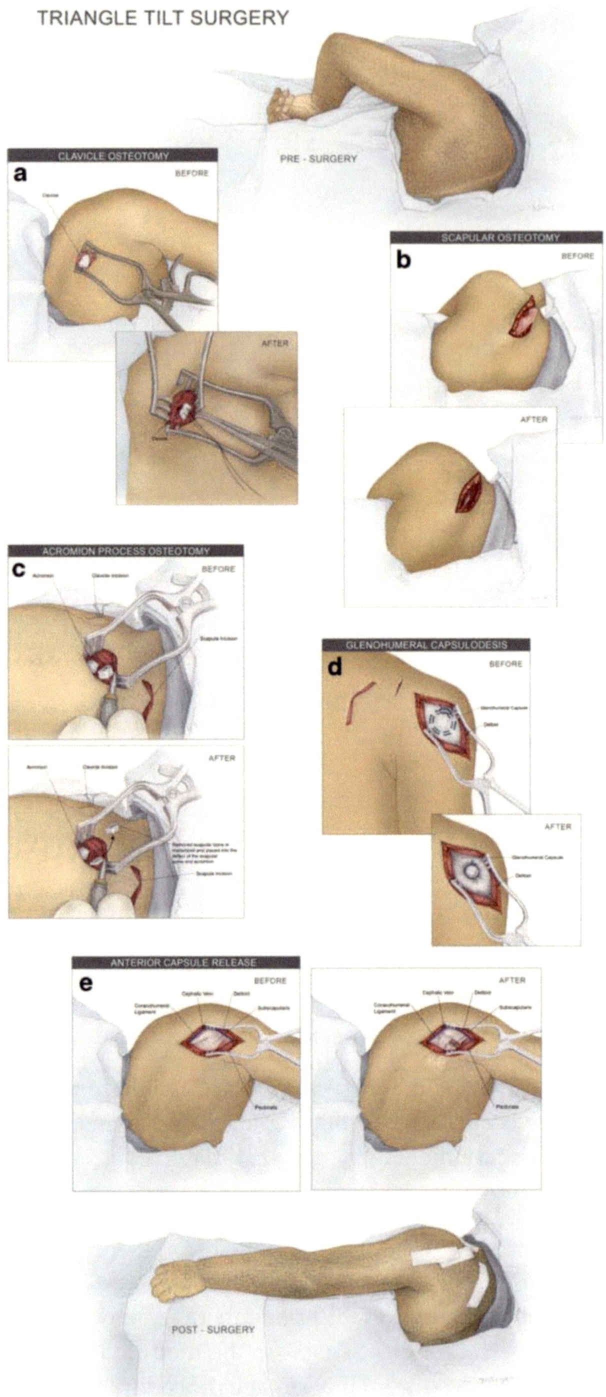

Figure 5. Artist's rendering of the triangle tilt surgery and anterior capsule release. Illustrated are osteotomies of the clavicle, scapula, and acromion process, along with glenohumeral capsulodesis and anterior capsule release. [19]

7. Conclusions

Based on our own experience with several thousand OBPI patients, we address primarily the muscle and bony operations, and we are less aggressive to nerve reconstruction. These procedures directly address the anatomy of the glenohumeral joint, and thereby resulting in the best possible overall functional outcome. In addition, this minimize the morbidity, expense and the invasiveness of surgery. Nerve reconstruction is reserved for those less common cases where the C5 and C6 nerve roots will not recover.

Modified Quad surgery improves median nerve conduction, and active abduction in young, as well as teen OBPI patients. The triangle tilt surgery improves all shoulder functions significantly in short (1 year), and long-term (2 years), and extended long term (5 years) follow-up, and further the functions are maintained over the extended time in these patients. We recommend that the triangle tilt surgery be performed before the age of 2 years for optimal improvements in outcomes of clinical functioning. This surgery can be performed as early as nine months, and up until late adolescence (16-17 years). Optimal clinical outcomes are achieved if this surgical procedure is performed before the age of 2 years, but improvements in functioning are observed if the surgery is performed after this age as well.

Author details

Rahul K. Nath* and Chandra Somasundaram

*Address all correspondence to: drnath@drnathmedical.com

Texas Nerve and Paralysis Institute, Houston, Texas, USA

References

[1] Adler JB, Patterson RL, Jr.: Erb's palsy. Long-term results of treatment in eighty-eight cases. *J Bone Joint Surg (Am)* 1967, 49:1052-1064.

[2] Hoeksma AF, Wolf H, Oei SL: Obstetrical brachial plexus injuries: incidence, natural course and shoulder contracture. *Clin Rehabil* 2000, 14:523-526.

[3] Foad SL, Mehlman CT, Ying J: The epidemiology of neonatal brachial plexus palsy in the United States. *J Bone Joint Surg (Am)* 2008, 90:1258-1264.

[4] Gurewitsch ED, Johnson E, Hamzehzadeh S, Allen RH: Risk factors for brachial plexus injury with and without shoulder dystocia. *Am J Obstet Gynecol* 2006, 194:486-492.

[5] Kay SP: Obstetrical brachial palsy. *Br J Plast Surg* 1998, 51:43-50.

[6] Zafeiriou DI, Psychogiou K: Obstetrical brachial plexus palsy. *Pediatr Neurol* 2008, 38:235-242.

[7] Akel BS, Oksuz C, Oskay D, Firat T, Tarakci E, Leblebicioglu G: Health-related quality of life in children with obstetrical brachial plexus palsy. *Qual Life Res* 2013.

[8] Royal College of Obstetricians and Gynaecologists: Guideline No. 42: Shoulder Dystocia. 2005.

[9] Jevitt CM: Shoulder dystocia: etiology, common risk factors, and management. *J Midwifery Womens Health* 2005, 50:485-497.

[10] Mahony R, Foley M, McAuliffe F, O'Herlihy C: Maternal weight characteristics influence recurrence of fetal macrosomia in women with normal glucose tolerance. *Aust N Z J Obstet Gynaecol* 2007, 47:399-401.

[11] Gherman RB, Chauhan S, Ouzounian JG, Lerner H, Gonik B, Goodwin TM: Shoulder dystocia: the unpreventable obstetric emergency with empiric management guidelines. *Am J Obstet Gynecol* 2006, 195:657-672.

[12] Rossi LN, Vassella F, Mumenthaler M: Obstetrical lesions of the brachial plexus. Natural history in 34 personal cases. *Eur Neurol* 1982, 21:1-7.

[13] Tada K, Tsuyuguchi Y, Kawai H: Birth palsy: natural recovery course and combined root avulsion. *J Pediatr Orthop* 1984, 4:279-284.

[14] Nath RK, Kumar N, Avila MB, Nath DK, Melcher SE, Eichhorn MG, Somasundaram C: Risk factors at birth for permanent obstetric brachial plexus injury and associated osseous deformities. *ISRN Pediatr*, 2012:307039.

[15] Al-Qattan MM: Obstetric brachial plexus palsy associated with breech delivery. *Ann Plast Surg* 2003, 51:257-264; discussion 265.

[16] Ubachs JM, Slooff AC, Peeters LL: Obstetric antecedents of surgically treated obstetric brachial plexus injuries. *Br J Obstet Gynaecol* 1995, 102:813-817.

[17] Gherman RB, Goodwin TM, Ouzounian JG, Miller DA, Paul RH: Brachial plexus palsy associated with cesarean section: an in utero injury? *Am J Obstet Gynecol* 1997, 177:1162-1164.

[18] Laurent JP, Lee RT: Birth-related upper brachial plexus injuries in infants: operative and nonoperative approaches. *J Child Neurol* 1994, 9:111-117; discussion 118.

[19] Nath RK: *Obstetric brachial plexus injuries-Erb's palsy: The Nath method of diagnosis and treatment.* College Station, TX: VirtualBookworm.com Publishing; 2007.

[20] Birch R, Bonney G, Wynn Parry CB: Birth lesions of the brachial plexus. In *Surgical disorders of the peripheral nerves.* Edited by Birch R, Bonney G, Wynn Parry CB. New York, NY: Churchill Livingstone; 1998: 209-233

[21] Gilbert A, Tassin JL: [Surgical repair of the brachial plexus in obstetric paralysis]. *Chirurgie* 1984, 110:70-75.

[22] Narakas AO: Obstetrical brachial plexus injuries. In *The paralysed hand. Volume* 2. Edited by Lamb DW. Edinburgh: Churchill-Livingstone; 1987: 116-135: *The Hand and upper limb].*

[23] Synek VM: Role of somatosensory evoked potentials in the diagnosis of peripheral nerve lesions: recent advances. *J Clin Neurophysiol* 1987, 4:55-73.

[24] Vredeveld JW, Blaauw G, Slooff BA, Richards R, Rozeman SC: The findings in paediatric obstetric brachial palsy differ from those in older patients: a suggested explanation. *Dev Med Child Neurol* 2000, 42:158-161.

[25] Colon AJ, Vredeveld JW, Blaauw G, Slooff AC, Richards R: Extensive somatosensory innervation in infants with obstetric brachial palsy. *Clin Anat* 2003, 16:25-29.

[26] Malessy MJ, Pondaag W, van Dijk JG: Electromyography, nerve action potential, and compound motor action potentials in obstetric brachial plexus lesions: validation in the absence of a "gold standard". *Neurosurgery* 2009, 65:A153-159.

[27] Malessy MJ, Pondaag W, Yang LJ, Hofstede-Buitenhuis SM, le Cessie S, van Dijk JG: Severe obstetric brachial plexus palsies can be identified at one month of age. *PLoS One* 2011, 6:e26193.

[28] Vanderhave KL, Bovid K, Alpert H, Chang KW, Quint DJ, Leonard JA, Jr., Yang LJ: Utility of electrodiagnostic testing and computed tomography myelography in the preoperative evaluation of neonatal brachial plexus palsy. *J Neurosurg Pediatr* 2012, 9:283-289.

[29] Friedman RJ, Hawthorne KB, Genez BM: The use of computerized tomography in the measurement of glenoid version. *J Bone Joint Surg (Am)* 1992, 74:1032-1037.

[30] Rollnik JD, Hierner R, Schubert M, Shen ZL, Johannes S, Troger M, Wohlfarth K, Berger AC, Dengler R: Botulinum toxin treatment of cocontractions after birth-related brachial plexus lesions. *Neurology* 2000, 55:112-114.

[31] Basciani M, Intiso D: Botulinum toxin type-A and plaster cast treatment in children with upper brachial plexus palsy. *Pediatr Rehabil* 2006, 9:165-170.

[32] Heise CO, Lorenzetti L, Marchese AJ, Gherpelli JL: Motor conduction studies for prognostic assessment of obstetrical plexopathy. *Muscle Nerve* 2004, 30:451-455.

[33] DeMatteo C, Bain JR, Galea V, Gjertsen D: Botulinum toxin as an adjunct to motor learning therapy and surgery for obstetrical brachial plexus injury. *Dev Med Child Neurol* 2006, 48:245-252.

[34] Birch R: Late sequelae at the shoulder in obstetrical palsy in children. In *Surgical techniques in orthopaedics and traumatology: Shoulder. Volume* 3. Edited by Randelli M,

Karlsson J. Paris: Elsevier; 2001: 55-200-E-210: *Surgical Techniques in Orthopaedics and Traumatology*].

[35] Kon DS, Darakjian AB, Pearl ML, Kosco AE: Glenohumeral deformity in children with internal rotation contractures secondary to brachial plexus birth palsy: intraoperative arthrographic classification. *Radiology* 2004, 231:791-795.

[36] van der Sluijs JA, van Ouwerkerk WJ, de Gast A, Wuisman PI, Nollet F, Manoliu RA: Deformities of the shoulder in infants younger than 12 months with an obstetric lesion of the brachial plexus. *J Bone Joint Surg (Br)* 2001, 83:551-555.

[37] Nath RK, Mackinnon SE, Jensen JN, Parks WC: Spatial pattern of type I collagen expression in injured peripheral nerve. *J Neurosurg* 1997, 86:866-870.

[38] Birch R: Invited editorial: Obstetric brachial plexus palsy. *J Hand Surg (Br)* 2002, 27 B: 3-8.

[39] Waters PM: Obstetric Brachial Plexus Injuries: Evaluation and Management. *J Am Acad Orthop Surg* 1997, 5:205-214.

[40] Nath RK, Paizi M: Scapular deformity in obstetric brachial plexus palsy: a new finding *Surg Radiol Anat* 2007, 29:133-140.

[41] Nath RK, Humphries AD: Computed tomography of the shoulders in patients with obstetric brachial plexus injuries: a retrospective study. *Ann Surg Innov Res* 2008, 2:4.

[42] Nath RK, Somasundaram C, Melcher SE, Bala M, Wentz MJ: Arm rotated medially with supination-the ARMS variant: description of its surgical correction. *BMC Musculoskelet Disord* 2009, 10:32.

[43] Shenaq SM, Kim JY, Armenta AH, Nath RK, Cheng E, Jedrysiak A: The Surgical Treatment of Obstetric Brachial Plexus Palsy. *Plast Reconstr Surg* 2004, 113:54E-67E.

[44] Nath RK, Somasundaram C, Mahmooduddin F: Comparing functional outcome of triangle tilt surgery performed before versus after two years of age. *Open Orthop J* 2011, 5:59-62.

[45] Waters PM, Bae DS: The early effects of tendon transfers and open capsulorrhaphy on glenohumeral deformity in brachial plexus birth palsy. Surgical technique. *J Bone Joint Surg Am* 2009, 91 Suppl 2:213-222.

[46] Waters PM, Bae DS: Effect of tendon transfers and extra-articular soft-tissue balancing on glenohumeral development in brachial plexus birth palsy. *J Bone Joint Surg (Am)* 2005, 87:320-325.

[47] Al-Qattan MM: Latissimus dorsi transfer for external rotation weakness of the shoulder in obstetric brachial plexus palsy. *J Hand Surg (Br)* 2003, 28:487-490.

[48] El-Gammal TA, Saleh WR, El-Sayed A, Kotb MM, Imam HM, Fathi NA: Tendon transfer around the shoulder in obstetric brachial plexus paralysis: clinical and computed tomographic study. *J Pediatr Orthop* 2006, 26:641-646.

[49] Pagnotta A, Haerle M, Gilbert A: Long-term results on abduction and external rotation of the shoulder after latissimus dorsi transfer for sequelae of obstetric palsy. *Clin Orthop Relat Res* 2004:199-205.

[50] Safoury Y: Muscle transfer for shoulder reconstruction in obstetrical brachial plexus lesions. *Handchir Mikrochir Plast Chir* 2005, 37:332-336.

[51] van der Sluijs JA, van Ouwerkerk WJ, de Gast A, Nollet F, Winters H, Wuisman PI: Treatment of internal rotation contracture of the shoulder in obstetric brachial plexus lesions by subscapular tendon lengthening and open reduction: early results and complications. *J Pediatr Orthop B* 2004, 13:218-224.

[52] Nath RK, Liu X: Nerve reconstruction in patients with obstetric brachial plexus injury results in worsening of glenohumeral deformity: a case-control study of 75 patients. *J Bone Joint Surg (Br)* 2009, 91-B:649-654.

[53] Nath RK, Paizi M: Improvement in abduction of the shoulder after reconstructive soft-tissue procedures in obstetric brachial plexus palsy. *J Bone Joint Surg (Br)* 2007, 89:620-626.

[54] Nath RK, Somasundaram C: Successful outcome of modified quad surgical procedure in preteen and teen patients with brachial plexus birth palsy. *Eplasty*, 12:e54.

[55] Nath RK, Kumar N, Somasundaram C: Modified Quad surgery significantly improves the median nerve conduction and functional outcomes in obstetric brachial plexus nerve injury. *Ann Surg Innov Res*, 7:5.

[56] Narakas AO: Muscle transpositions in the shoulder and upper arm for sequelae of brachial plexus palsy. *Clin Neurol Neurosurg* 1993, 95 Suppl:S89-91.

[57] Nath RK, Karicherla P, Mahmooduddin F: Shoulder function and anatomy in complete obstetric brachial plexus palsy: long-term mprovement after triangle tilt surgery. *Child's Nervous System* 2010, 26:1009-1019.

[58] Nath RK, Avila MB, Karicherla P: Triangle tilt surgery as salvage procedure for failed shoulder surgery in obstetric brachial plexus injury. *Pediatr Surg Int* 2010, 26:913-918..

[59] Nath RK, Amrani A, Melcher SE, Eichhorn MG: Triangle tilt surgery in an older pediatric patient with obstetric brachial plexus injury. *ePlasty* 2009, 9:e26.

[60] Nath RK, Amrani A, Melcher SE, Wentz MJ, Paizi M: Surgical normalization of the shoulder joint in obstetric brachial plexus injury. *Ann Plast Surg* 2010, 65:411-417.

[61] Nath RK, Avila MB, Karicherla P, Somasundaram C: Assessment of triangle tilt surgery in children with obstetric brachial plexus injury using the pediatric outcomes data collection instrument. *Open Orthop J* 2011, 5:385-388.

[62] Nath RK, Liu X, Melcher SE, Fan J: Long-term outcomes of triangle tilt surgery for obstetric brachial plexus injury. *Pediatr Surg Int* 2010, 26:393-399.

[63] Nath RK, Mahmooduddin F: Triangle tilt surgery: effect on coracohumeral distance and external rotation of the glenohumeral joint. *Eplasty* 2010, 10:e67.

[64] Nath RK, Melcher SE, Lyons AB, Paizi M: Surgical correction of the medial rotation contracture in obstetric brachial plexus palsy. *J Bone Joint Surg (Br)* 2007, 89:1638-1644.

[65] Nath RK, Melcher SE, Paizi M: Surgical correction of unsuccessful derotational humeral osteotomy in obstetric brachial plexus palsy: Evidence of the significance of scapular deformity in the pathophysiology of the medial rotation contracture. *J Brachial Plex Peripher Nerve Inj* 2006, 1:9.

[66] Nath RK, Somasundaram C, Mahmooduddin F: Triangle tilt and steel osteotomy: similar approaches to common problems. Open Orthop J 2011, 5:124-133.

Chapter 5

A Current Overview of Diabetic Neuropathy – Mechanisms, Symptoms, Diagnosis, and Treatment

Takashi Kawano

Additional information is available at the end of the chapter

http://dx.doi.org/10.5772/58308

1. Introduction

Diabetic neuropathies are nerve disorders associated with diabetes, which affect approximately half of all diabetes patients [1]. The most common complication of diabetes is caused by hyperglycemia which can damage nerve fibers throughout the body [2]. Depending on the types of nerves involved, diabetic neuropathies can be categorized as peripheral, autonomic, proximal, focal neuropathies [3].

Because the pathogenesis mechanisms of diabetic neuropathy remain unknown, numerous studies try to elucidate the underlying mechanisms of this disease. Several reports have demonstrated that a variety of molecules are likely involved in the development of diabetic neuropathy, such as protein kinase C, polyol, aldose reductase, advanced glycation end-products, reactive oxygen species, cytokines [1-10]. Moreover, some risk factors including metabolite, autoimmune, inherited traits and lifestyle, may contribute to the development of diabetic neuropathy.

These multiple factors mentioned above might correlate with various symptoms of diabetic neuropathy. These symptoms vary in different organ systems, such as the extremities, digestive system, urinary tract, blood vessels, heart, and sex organs, depending on the nerves affected [9, 10]. The symptoms usually include pain, foot ulcer, dysesthesia, numbness and tingling of extremities, indigestion, nausea, vomiting, diarrhea, facial and eyelid drooping, eyesight change, dizziness, muscle weakness, dysphagia, urinary incontinence, sexual dysfunction, and speech impairment [2, 4, 9-11]

The symptoms remain minor initially and develop gradually over years. As a result, the majority of patients do not even realize they are affected until the complications become noticeable or severe. Accordingly, it is difficult to diagnose the disease in the early stages. However, doctors can diagnose diabetic neuropathy based on the patients' symptoms and physical examinations usually including ankle reflexes, loss of sensation in the extremities, blood pressure, heart rate, muscle strength, vibration, temperature, or light touch [11-12]. In addition, nerve conduction test, electromyography and ultrasound test may help diagnose the disease [3, 4].

Due to the poorly understood mechanism, effective therapies that can cure diabetic neuropathy remain elusive. However, there exist various options to prevent or treat the disease. To date, the fundamental treatment for diabetic neuropathy is to keep blood glucose levels under control to prevent further nerve damage [4]. Additionally, drug treatment also helps relieve pain and other symptoms. The medications include tricyclic antidepressants, classic analgesics, serotonin reuptake inhibitors and antiepileptic drugs [3, 13].

Because of the side effects of drug therapy, physical treatment can help alleviate pain and some other symptoms, such as foot ulcer, muscle weakness, loss of sensation and sexual dysfunction. The physical treatment include electrical nerve stimulation, gait training, posture training, manual therapy, exercise programs, foot care, therapeutic ultrasound, hot wax, short wave diathermy, photo energy therapy [12, 14, 15]. Moreover, healthy lifestyle, quitting smoking will be beneficial to diabetic neuropathy. Recently, cell therapy has been proposed to treat diabetic neuropathy [16].

In this chapter, we will discuss the mechanisms, symptoms, diagnosis, and treatment of diabetic neuropathy.

2. Epidemiology

The incidence of diabetic neuropathy is the highest among diabetic complications, and diabetic neuropathy develops early after the onset of diabetes [1, 13, 17]. The risk factors of diabetic neuropathy are hyperglycemia and its persistence (Table 1). Hypertension, dyslipidemia, obesity, and cigarette smoking are also included in the risk factors in Western countries [1, 13, 17].

Factors	**Odds ratio**
Total cholesterol	1.15
Triglyceride	1.21
Body mass index	1.27
HbA1c change degree	1.36
Smoking	1.38
Duration of diabetes mellitus	1.40
HbA1c level	1.48
Hypertension	1.57

Adjusted odds ratio for associations between key risk factors and the incidence of diabetic neuropathy with logistic regression model; HbA1c: Hemoglobin A1c

Table 1. Risk factors of diabetic neuropathy [21, 22]

For the prevention of diabetic neuropathy, blood glucose control is the most important [18, 19]. In a study investigating the prevalence of diabetic neuropathy in diabetic patients and whether patients recognized the development of neuropathy, clinical diabetic neuropathy was noted in 14% on average but not recognized by most patients [20].

3. Pathological mechanism

The pathological mechanism of diabetic neuropathy cannot be explained with a single cause, and various hypotheses have been proposed (Table 2). These are roughly divided into metabolic [23], vascular [24], and neuroregeneration disorder hypotheses [25].

1. Activation of polyol pathway
2. Down-regulation of intracellular myoinositol
3. Dysfunction of protein kinase C
4. Down-regulation of intracellular cyclic AMP
5. Inhibition of $Na^+/K^+/ATPase$
6. Degradation of nitric oxide
7. Advance of protein glycation
8. Increase of free radical
9. Disorder of polyunsaturated fatty acid synthesis
10. Disorder of prostaglandin synthesis
11. Action attenuation of a nerve growth factor
12. Nerve blood flow degradation, nerve vascular resistance enhancement
AMP: Adenosine monophosphate

Table 2. Potential pathogenesis of diabetic neuropathy

3.1. Impairment of polyol pathway

Altered peripheral nerve polyol metabolism has been implicated as a central factor in the pathogenesis of diabetic neuropathy. Aldose reductase converts glucose to sorbitol (such as polyol) using nicotinamide adenine dinucleotide phosphate (NADPH) as a coenzyme (Figure 1). Sorbitol is further converted to fructose by sorbitol dehydrogenase using nicotinamide adenine dinucleotide (NAD^+) as a coenzyme, constituting the bypass polyol pathway of glucose metabolism [26].

In hyperglycemia accompanying diabetes, the cellular glucose level rises independently from insulin, resulting in enhancement of aldose reductase activity, which elevates the intracellular sorbitol level and, subsequently, the intracellular osmotic pressure. This condition induces functional and structural abnormalities in tissue and cells.

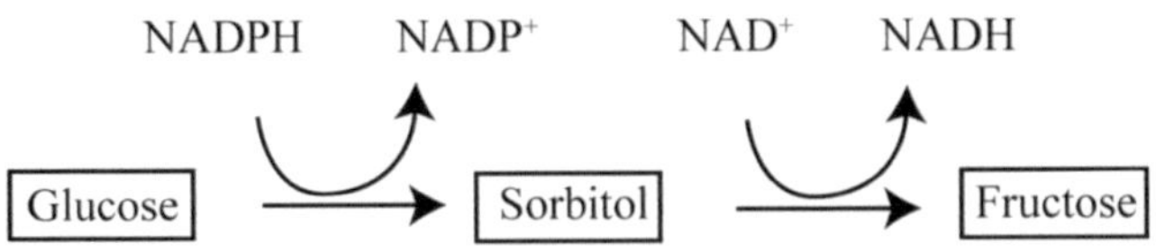

Figure 1. Polyol pathway. The polyol pathway consists of two-step metabolic pathway.

An aldose reductase reduces glucose in sorbitol. This reaction oxidizes nicotinamide adenine dinucleotide phosphate (NADPH) to $NADP^+$(the oxidized form of NADPH). Subsequently, sorbitol dehydrogenase enzymatically oxidizes sorbitol to fructose, which also produces nicotinamide adenine dinucleotide (NADH) from nicotinamide adenine dinucleotide (NAD $^+$). The inhibition of the aldose reductase is one of key element in the prevention of diabetic complications.

In addition to osmotic pressure elevation, sorbitol accumulation decreases the intracellular myoinositol content, which inhibits phosphoinositide metabolism and reduces protein kinase C and Na^+/K^+/ATPase activities in peripheral nerves, being involved in the manifestation of diabetic neuropathy.

3.2. Activation of protein kinase C

Hyperglycemia promotes the synthesis of an endogenous protein kinase C activator, diacylglycerol [27-30]. Actually, excess activation of β2-type protein kinase C in cardiovascular tissue in an animal diabetes model has been reported. Enhanced vascular protein kinase C is involved in permeability, the contractile force, and the differentiation and proliferation of cells.

Excess protein kinase C activation induces ischemia in peripheral nerves through increased vascular permeability and thickening of the basement membrane and causes neuropathy.

3.3. Increase in oxidative stress

Hyperglycemia enhances NADPH oxidase expression and the endothelial nitric oxide synthase (eNOS) uncoupling reaction in vascular endothelial cells, through which superoxide is excessively produced [4, 31-33]. Nitric oxide (NO) is essential for endothelial cell function.

Excess superoxide decreases NO by binding to it, and this binding reaction promotes the secondary synthesis of reactive oxygen species (ROS), such as peroxynitrite and hydroxyl radicals. ROS have strong cytotoxicity, and an increase in ROS induces neurosis.

3.4. Other factors

Bone marrow-derived proinsulin-and tumor necrosis factor-α (TNFα)-producing cells appear in a diabetic state [5, 34, 35]. These cells enter the dorsal root ganglions and peripheral nerves

http://dx.doi.org/10.5772/58308

(axon and Schwann cells) and induce cell fusion. Fused cells impair Ca^{2+}homeostasis and induce apoptosis. The appearance of these abnormal cells is resolved by insulin treatment.

It has also been clarified that the abnormality of intracellular signal transmission systems in nerve tissues including that of insulin signals is closely involved in abnormal peripheral nerve function [36]. The peripheral neuropathy developmental mechanism may be a new target of neuropathy treatment, other than blood glucose control.

4. Symptoms

The manifestation of subjective symptoms of diabetic neuropathy is the earliest among complications of diabetic patients, and the incidence is the highest [1, 13, 17, 37]. Its pathology starts with numbness and sensory disturbance of the four limbs, and manifests various clinical pictures, such as autonomic neuropathy and mononeuropathy (Table 3).

1. Sensory disturbance is dominant
2. A disorder of an inferior limb is dominant, and a disorder of a superior limb is mild
3. Vibratory sensation is disordered since early stage
4. A tendon reflex of an inferior limb decreases since early stage
5. Ophthalmoplegia often accompanies
6. Autonomic neuropathy often accompanies

Table 3. Clinical features of diabetic neuropathy

Sensory symptoms accompanying diabetic neuropathy, such as pain and numbness, distress patients, and subsequent hypoesthesia leads to the primary cause of lower limb amputation, diabetic gangrene [9, 10, 38, 39]. Diverse symptoms of autonomic neuropathy (Table 4) markedly reduce the Quality of Life (QOL) of patients [40, 41, 42].

1. Constipation, diarrhea, gastric hypokinesia (dull feeling in the stomach)
2. Dizziness (orthostatic hypotension)
3. Silent myocardial infarction: Myocardial infarction or angina without chest pain
4. Dysuria
5. Erectile dysfunction
6. Non-symptomatic hypoglycemia

Table 4. Diabetic autonomic neuropathy

Clinically, there are several disease types of diabetic neuropathy based on the distribution of disorders and developmental pattern (Table 5).

1. Hyperglycemic neuropathy
2. Symmetric polyneuropathy
1) Sensory / autonomic neuropathy
2) Acute painful diabetic neuropathy
3. Focal and multifocal neuropathy
1) Cranial neuropathy
2) Thoraco-abdominal neuropathy
3) Focal limb neuropathy
4) Diabetic amyotrophy
4. Mixed forms

Table 5. Classification of diabetic neuropathy [43]

In diabetic neuropathy, sensory neuropathy is dominant, but subjective sensory symptoms generally do not extend to the proximity from the ankle joint in many cases, and its onset is associated with numbness and pain of the toes and sole. The fingers are asymptomatic in this stage, showing "tabi (socks with the big toe separated)-type" sensory symptoms, and this pattern is frequently noted in routine medical practice.

In the late stage, "glove-socks-type" sensory abnormality manifests. Diabetic neuropathy cases with the expansion of sensory symptoms to the precordium and parietal region have been reported. This neurologic manifestation pattern is derived from the advancement pattern of axon degeneration, and it occurs because the nerves in the lower limbs are longer than those in the upper limbs.

Since diabetic neuropathy progresses slowly, the divergence between the upper and lower limb symptoms may continue for a relatively long time. Regarding sensory disturbance, in diabetic neuropathy in which positive symptoms of the feet, such as numbness and pain, develop in the early to middle stage and negative symptoms, such as hypoesthesia, develop in the terminal stage, generally, an abnormal autonomic nerve function appears from the early stage and then autonomic nerve symptoms may manifest, but the manifestation of motor neuropathy is late (Table 6).

N0		no neuropathy
N1		Asymptomatic neuropathy
	N1a	Abnormal of examination without neuropathy symptom
	N1b	Abnormal of examination with neurologic signs without neuropathy symptom
N2		Symptomatic neuropathy
	N2a	Abnormal of examination with neurologic signs with neuropathy symptom
	N2b	N2a plus weakness of ankle dorsiflexion
N3		Disabling neuropathy

Table 6. Severity grade of diabetic neuropathy [3]

5. Diagnosis

Diabetic neuropathy can be diagnosed when the patient has been diagnosed with diabetes and other diseases causing polyneuropathy have been ruled out. Diseases required to be differentiated are shown in Table 7.

There are no diabetic neuropathy-specific symptoms or tests, and no diagnostic criteria with international consensus have been established. Diabetic neuropathy has to be comprehensively diagnosed based on various neurologic manifestations and test results [44-46].

The symptom characteristic of diabetic neuropathy is bilateral symmetric polyneuropathy with dominance on the distal side, and it more frequently develops from the lower limbs, particularly from the feet and crura, than from the upper limbs.

1. Ongoing diabetes mellitus
2. There is no disorder to cause neurological symptom besides diabetes mellitus
3. Symmetric symptom (spontaneous pain, paresthesia, hypaesthesia, anesthesia)
4. Attenuation of reflexes in the ankle or knee
5. Pallesthesia
6. Abnormal of electrophysiological neurologic function tests
7. Symptoms of autonomic neuropathy

Table 7. Diagnosis of diabetic neuropathy

Subjective symptoms are an abnormal sensation, cold sense, and hypoesthesia of the feet. When thick myelinated nerve fibers are mainly impaired, an increase in the pallesthesia threshold and reduction/loss of tactile sensation of the toes, movement velocity, sensory nerve conduction velocity, and the tendon reflex are observed. When thin nerve fibers and unmyelinated nerves are impaired, an increase in the thermal sensation threshold and features of autonomic neuropathy are observed. When 3 or more of these 4 items are present, the patient is diagnosed with diabetic peripheral neuropathy.

The peripheral neuropathy signs important to objectively diagnose the disease stage of diabetic neuropathy are summarized below:

5.1. Reduction/loss of Achilles tendon reflex

Since this symptom is frequently observed even in patients showing no symptoms, it is very important to identify diabetic neuropathy in the asymptomatic stage [2, 4, 9-11].

A test in a kneeling posture (Babinski position), in which loss of the reflex can be readily observed, is recommended. Many cases of diabetic neuropathy show bilateral abnormality, and apparent laterality is a sign of lumbar vertebral disease [47].

5.2. Pallesthesia

The impairment of vibration perception threshold is used to early diagnosis of peripheral neuropathy [48-50].

An aluminum 128-Hz tuning folk is standard for the examination of pallesthesia. Since the vibration of a tuning folk exponentially attenuates, the time required to reach the threshold is almost constant when it is hit with a force stronger than a specific level. The base of a vibrating tuning fork was placed on the hallux of the patient. The examiner asks the patients first if the vibration is perceived. Next, the patient should inform the examiner when the vibration stops. The diagnosis of diabetic neuropathy is to be suspected if the vibration duration sensation is less than 10 seconds.

5.3. Peripheral nerve conduction velocity test

In this test, peripheral nerves are stimulated with electricity through the skin, and the nerve conduction velocity and waveform are analyzed based on the reactions to diagnose and treat diseases. When neuropathy occurs, the nerve conduction velocity decreases [51-53].

5.4. Monofilament

Activity of nerves perceiving tactile and pressure sensations is investigated by attaching a monofilament to the foot. Perception decreases in diabetic neuropathy patients [54, 55].

5.5. Coefficient of respiratory heart rate variability

This is an autonomic nerve function test. Variation in the pulse with deep breaths compared to that on rest is investigated using electrocardiography. Normally, pulse variation increases on deep breathing, but this variation decreases when autonomic nerves are impaired [56].

6. Treatment

Early-stage diabetic neuropathy can be improved by blood glucose control alone, but it becomes intractable after progression to a certain stage. Aldose reductase inhibitors are being developed for treatment based on the metabolic disorder hypothesis of diabetic neuropathy, but treatment with these drugs alone may be insufficient [57].

6.1. Blood glucose control

In a large-scale intervention study, Diabetes Control and Complications Trial (DCCT; http://diabetes.niddk.nih.gov/dm/pubs/control/), 1,441 patients with insulin-dependent diabetes received intensive insulin therapy or conventional insulin treatment for 6.5 years on average [58]. In the intensive insulin therapy group, significant inhibition of the development and advancement of neuropathy was demonstrated, showing that strict blood glucose control is important for the prevention and treatment of diabetic neuropathy. However, rapid blood glucose control exacerbates neuropathy in some patients, and this condition is termed post-treatment neuropathy. In these patients, neuropathy may have been present before the initiation of blood glucose control. Generally, pain remits within one year. Thus, it is important to relieve patients and remove their anxiety. For patients with poor blood glucose control and complications, it is safe to slowly control blood glucose.

6.2. Aldose reductase inhibitor

Aldose reductase inhibitor inhibits the enhancement of polyol metabolic activity, a mechanism of diabetic neuropathy development, and it is expected to be a specific therapeutic drug for diabetic neuropathy [59-61].

Many aldose reductase inhibitors have been developed, and clinical efficacy was noted in some. However, the evidence for the efficacy of aldose reductase inhibitor for diabetic neuropathy is still insufficient. Epalrestat is a typical aldose reductase inhibitor. In a multi-center controlled clinical study with this drug, the conduction velocity of the median nerve decreased over years in the untreated group, but the drug inhibited it. The effect was marked in patients with favorable blood glucose control and a short duration of diabetic neuropathy. Thus, it is desirable to administer epalrestat in consideration of the indication. The possibility of epalrestat improving the autonomic nerve function has been reported, although it was a small-scale study [59].

6.3. Antioxidants

The usefulness of antioxidants has been tested with regard to abnormal protein kinase C (PKC) activity and oxidative stress, and the improvement of neurologic manifestations and physical findings by α-lipoic acid has been reported [62, 63].

6.4. Incretin

Incretin (glucagon-like peptide-1: GLP-1 and glucose-dependent insulinotropic polypeptide: GIP) has recently been attracting attention as a new anti-diabetes drug [64, 65].

Incretin has also been shown to act on cells or tissues other than pancreatic β cells, i.e., extrapancreatic actions [66]. Medical-experimentally, incretin and related drugs have various neuroprotective actions, and the possibility of incretin being effective for diabetic neuropathy has been reported [64, 65, 67].

6.5. Regeneration therapy

Functional improvement of vascular and nerve cells and regeneration of degenerated tissue corresponding to the pathology of diabetic neuropathy are expected radical treatments of diabetic neuropathy [16, 68].

In studies on regenerative medicine for diabetic neuropathy, precursor and stem cells isolated and cultured from the bone marrow and fat tissue, stem cells induced to differentiate from embryonic stem (ES) and induced pluripotent stem (iPS) cells, and bone marrow mononuclear cells containing many of these precursor and stem cells are mainly used. Further investigation aiming at clinical application is necessary.

6.6. Others

For the improvement of blood flow, prostaglandin E_1, an oral prostacyclin derivative, cilostazol, and eicosapentaenoic acid (EPA) are effective in some cases.

6.7. Symptomatic treatment of pain

Pain develops in most disease types of diabetic neuropathy [69, 70].

Drug	Action
Tricyclic antidepressant	Serotonin–norepinephrine reuptake inhibitor
Carbamazepine	Na^+ channel block
Valproate	Central inhibition via augmentation of GABA
Topiramate	Na^+ channel and AMPA receptor block
Lamotrigine	Na^+ channel block, central inhibition
Dextromethorphan	Glutamate N-methyl-D-aspartate receptor antagonists
Tramadol	Weak μ-opioid receptor agonist, Serotonin–norepinephrine reuptake inhibitor
Mexiletine	Na^+ channel block
Capsaicin	Activation of transient receptor potential cation channel subfamily V member 1
Gabapentin	α2δ Ca^{2+} channel inhibition
Pregabalin	α2δ Ca^{2+} channel inhibition
Duloxetine	Serotonin–norepinephrine reuptake inhibitor

GABA: gamma-aminobutyric acid, AMPA: α-amino-3-hydroxy-5-methyl-4-isoxazolepropionic acid

Table 8. Drugs currently used in treatment of diabetic neuropathy and its action

Although the developmental mechanism of pain has not been fully clarified, the activation of $Na^+$and Ca^{2+}channels in peripheral nerves is closely involved, and mexiletine and anticonvulsants, with inhibitory actions, are effective.

The involvement of activation on the central side including the posterior horn of the spinal cord increases as the condition becomes chronic, and tricyclic antidepressants, selective serotonin reuptake inhibitor (SSRI), serotonin–norepinephrine reuptake inhibitor (SNRI), and N-methyl-D-aspartate (MNDA) receptor antagonists, have become important with regard to the site of action. The efficacy of opioids (tramadol and oxycodone) has also been reported.

On meta-analysis, tricyclic antidepressants were most effective. Among anticonvulsants, the conventional type (carbamazepine and phenytoin) has been reported to be superior to the new

http://dx.doi.org/10.5772/58308

type (gabapetine and pregabalin), with regard to the efficacy and adverse effects. Capsaicin and lidocaine patches are also useful to alleviate symptoms.

6.8. Treatment of autonomic neuropathy

When autonomic neuropathy appears, organs innervated by autonomic nerves become functionally abnormal, and diverse symptoms develop, such as dyshidrosis, orthostatic hypotension, gastric asthenia, stool abnormality, bladder and erectile dysfunctions, and hypoglycemia unawareness. When neuropathy is mild, modification of the blood glucose control and lifestyle improves these functional disorders in many cases. When neuropathy is advanced and impairs daily living activities, symptomatic treatment with drugs corresponding to the symptoms is necessary [41, 71].

For orthostatic hypotension, firstly, drugs likely to decrease the blood pressure are withdrawn, and patients are instructed to avoid rapid postural changes while standing. Frequent ingestion of a small amount of food is effective to prevent postprandial blood pressure reduction. Compression of the lower limbs and abdominal region by wearing elastic underwear is effective for orthostatic hypotension. Salt ingestion and the administration of fludrocortisone acetate are also effective, but these are likely to cause edema and heart failure, to which attention should be paid.

For erectile dysfunction, firstly, drugs likely to cause it should be withdrawn. For patients requiring drug therapy, a phosphodiesterase inhibitor, sildenafil or vardenafil, is effective. However, these are contraindicated for patients being treated with nitroglycerin and nitrous acid medicine for ischemic heart disease because a phosphodiesterase inhibitor is very likely to cause serious blood pressure reduction.

Gastric asthenia is treated with the frequent ingestion of a small amount of food and restriction of fat and fiber ingestion. Symptoms are improved by these symptomatic treatments alone in many mild cases. When drug therapy is necessary, metoclopramide and domperidone are effective, but long-term administration may induce extrapyramidal symptoms as adverse effects, to which attention should be paid.

7. Conclusion

Diabetic neuropathy is caused by dysfunction of the peripheral or central nervous system associated with abnormally high levels of blood glucose. It is often chronic and disabling. Advanced neuropathy not only reduces QOL of patients but also influences their vital prognosis, shown by the high mortality of patients with autonomic neuropathy. Therefore, to improve the vital prognosis and QOL of patients, it is important to perform periodic neurological examination from the early stage for the early diagnosis and treatment of diabetic neuropathy.

Author details

Takashi Kawano*

Department of Anesthesiology and Intensive Care Medicine, Kochi Medical School, Japan

References

[1] Hinder, LM., Vincent, AM., Burant, CF., Pennathur, S. & Feldman, EL. Bioenergetics in diabetic neuropathy: what we need to know. J Peripher Nerv Syst 2012; 17(Suppl. 2) 10-4.

[2] Yagihashi S, Mizukami H, Sugimoto K. Mechanism of diabetic neuropathy: Where are we now and where to go? Journal of Diabetes Investigation 2011; 2(1) 18–32.

[3] Tesfaye S, Boulton AJ, Dyck PJ, Freeman R, Horowitz M, Kempler P, Lauria G, Malik RA, Spallone V, Vinik A, Bernardi L, Valensi P: Toronto Diabetic Neuropathy Expert Group. Diabetic neuropathies: update on definitions, diagnostic criteria, estimation of severity, and treatments. Diabetes Care 2010; 33(10) 2285-93.

[4] Vincent AM, Russell JW, Low P, Feldman EL. Oxidative stress in the pathogenesis of diabetic neuropathy. Endocr Rev 2004; 25(4) 612-28.

[5] Chan L, Terashima T, Urabe H, Lin F, Kojima H. Pathogenesis of diabetic neuropathy: bad to the bone. Ann N Y Acad Sci 2011; 1240 70-6.

[6] Jack M, Wright D. Role of advanced glycation endproducts and glyoxalase I in diabetic peripheral sensory neuropathy. Transl Res 2012; 159(5) 355-65.

[7] Lorenzi M. The polyol pathway as a mechanism for diabetic retinopathy: attractive, elusive, and resilient. Exp Diabetes Res 2007; No.61038.

[8] Xia P, Kramer RM, King GL. Identification of the mechanism for the inhibition of Na^+, K^+-adenosine triphosphatase by hyperglycemia involving activation of protein kinase C and cytosolic phospholipase A2. J Clin Invest 1995; 96(2) 733-40.

[9] Dinh T, Tecilazich F, Kafanas A, Doupis J, Gnardellis C, Leal E, Tellechea A, Pradhan L, Lyons TE, Giurini JM, Veves A. Mechanisms involved in the development and healing of diabetic foot ulceration. Diabetes 2012; 61(11) 2937-47.

[10] Bagyánszki M, Bódi N. Diabetes-related alterations in the enteric nervous system and its microenvironment. World J Diabetes 2012; 3(5) 80-93.

[11] Al-Geffari M. Comparison of different screening tests for diagnosis of diabetic peripheral neuropathy in Primary Health Care setting. Int J Health Sci (Qassim) 2012; 6(2) 127-34.

[12] Balbinot LF, Canani LH, Robinson CC, Achaval M, Zaro MA. Plantar thermography is useful in the early diagnosis of diabetic neuropathy. Clinics (Sao Paulo) 2012; 67(12) 1419-25.

[13] Boulton AJ, Malik RA, Arezzo JC, Sosenko JM. Diabetic somatic neuropathies. Diabetes Care 2004; 27(6) 1458-86.

[14] Höke A. Animal models of peripheral neuropathies. Neurotherapeutics. 2012; 9(2) 262-9.

[15] Pieber K, Herceg M, Paternostro-Sluga T. Electrotherapy for the treatment of painful diabetic peripheral neuropathy: a review. J Rehabil Med 2010; 42(4) 289-95.

[16] Han JW, Sin MY, Yoon YS. Cell therapy for diabetic neuropathy using adult stem or progenitor cells. Diabetes Metab J 2013; 37(2) 91-105.

[17] Said G. Diabetic neuropathy—a review. Nat Clin Pract Neurol 2007; 3(6) (June), pp. 331-40.

[18] Dyck PJ, Davies JL, Clark VM, Litchy WJ, Dyck PJ, Klein CJ, Rizza RA, Pach JM, Klein R, Larson TS, Melton LJ 3rd, O'Brien PC. Modeling chronic glycemic exposure variables as correlates and predictors of microvascular complications of diabetes. Diabetes Care 2006; 29(10) 2282-8.

[19] Martin CL, Albers J, Herman WH, Cleary P, Waberski B, Greene DA, Stevens MJ, Feldman EL: DCCT/EDIC Research Group. Neuropathy among the diabetes control and complications trial cohort 8 years after trial completion. Diabetes Care 2006; 29(2) 340-4.

[20] Bongaerts BW, Rathmann W, Heier M, Kowall B, Herder C, Stöckl D, Meisinger C, Ziegler D. Older subjects with diabetes and prediabetes are frequently unaware of having distal sensorimotor polyneuropathy: the KORA F4 study. Diabetes Care 2013; 35(5) 1141-6.

[21] Tesfaye S, Chaturvedi N, Eaton SE, Ward JD, Manes C, Ionescu-Tirgoviste C, Witte DR, Fuller JH; EURODIAB Prospective Complications Study Group. Vascular risk factors and diabetic neuropathy. N Engl J Med 2005; 352(4) 341-50.

[22] Forsblom CM, Sane T, Groop PH, Tötterman KJ, Kallio M, Saloranta C, Laasonen L, Summanen P, Lepäntalo M, Laatikainen L, Matikainen E, Teppo AM, Koskimies S, Groop, L. Risk factors for mortality in Type II (non-insulin-dependent) diabetes: evidence of a role for neuropathy and a protective effect of HLA-DR4. Diabetologia 1998; 41(11) 1253-62.

[23] Zochodne DW. Diabetic polyneuropathy: an update. Curr Opin Neurol 2008; 21(5) 527-33.

[24] Dyck PJ. Hypoxic neuropathy: does hypoxia play a role in diabetic neuropathy? The 1988 Robert Wartenberg lecture. Neurology 1989; 39(1) 111-8.

[25] Yasuda H, Terada M, Maeda K, Kogawa S, Sanada M, Haneda M, Kashiwagi A, Kikkawa R. Diabetic neuropathy and nerve regeneration. Prog Neurobiol 2003; 69(4) 229-85.

[26] Yabe-Nishimura C. Aldose reductase in glucose toxicity: a potential target for the prevention of diabetic complications. Pharmacol Rev 1998; 50(1) 21-33.

[27] Borghini I, Ania-Lahuerta A, Regazzi R, Ferrari G, Gjinovci A, Wollheim CB, Pralong WF. Alpha, beta I, beta II, delta, and epsilon protein kinase C isoforms and compound activity in the sciatic nerve of normal and diabetic rats. J Neurochem 1994; 62(2) 686-96.

[28] Hempel A, Maasch C, Heintze U, Lindschau C, Dietz R, Luft FC, Haller H. High glucose concentrations increase endothelial cell permeability via activation of protein kinase C alpha. Circ Res 1997; 81(3) 363-71.

[29] Roberts RE, McLean WG. Protein kinase C isozyme expression in sciatic nerves and spinal cords of experimentally diabetic rats. Brain Res 1997; 754(1-2) 147-56.

[30] Geraldes P, King GL. Activation of protein kinase C isoforms and its impact on diabetic complications. Circ Res 2010; 106(8) 1319-31.

[31] Yorek MA. The role of oxidative stress in diabetic vascular and neural disease. Free Radic Res 2003; 37(5) 471-80.

[32] Pop-Busui R, Sima A, Stevens M. Diabetic neuropathy and oxidative stress. Diabetes Metab Res Rev 2006; 22(4) 257-73.

[33] Figueroa-Romero C, Sadidi M, Feldman EL. Mechanisms of disease: the oxidative stress theory of diabetic neuropathy. Rev Endocr Metab Disord 2008; 9(4) 301-14.

[34] Terashima T, Kojima H, Fujimiya M, Matsumura K, Oi J, Hara M, Kashiwagi A, Kimura H, Yasuda H, Chan L. The fusion of bone-marrow-derived proinsulin-expressing cells with nerve cells underlies diabetic neuropathy. Proc Natl Acad Sci U S A 2005; 102(35) 12525-30.

[35] Terashima T, Kojima H, Chan L. Bone marrow expression of poly(ADP-ribose) polymerase underlies diabetic neuropathy via hematopoietic-neuronal cell fusion. FASEB J 2012; 26(1) 295-308.

[36] Brussee V, Cunningham FA, Zochodne DW. Direct insulin signaling of neurons reverses diabetic neuropathy. Diabetes 2004; 53(7) 1824-30.

[37] Rutkove SB. A 52-year-old woman with disabling peripheral neuropathy: review of diabetic polyneuropathy. JAMA 2009; 302(13) 1451-8.

[38] Cappellari A, Airaghi L, Capra R, Ciammola A, Branchi A, Levi Minzi G, Bresolin N. Early peripheral nerve abnormalities in impaired glucose tolerance. Electromyogr Clin Neurophysiol 2005; 45(4) 241-4.

[39] Isak B, Oflazoglu B, Tanridag T, Yitmen I, Us O. Evaluation of peripheral and autonomic neuropathy among patients with newly diagnosed impaired glucose tolerance. Diabetes Metab Res Rev 2008; 24(7) 563-9.

[40] Vinik AI, Maser RE, Mitchell BD, Freeman R. Diabetic autonomic neuropathy. Diabetes Care 2003; 26(5) 1553-79.

[41] Low PA, Benrud-Larson LM, Sletten DM, Opfer-Gehrking TL, Weigand SD, O'Brien PC, Suarez GA, Dyck PJ. Autonomic symptoms and diabetic neuropathy: a population-based study. Diabetes Care 2004; 27(12) 2942-7.

[42] Zilliox L, Peltier AC, Wren PA, Anderson A, Smith AG, Singleton JR, Feldman EL, Alexander NB, Russell JW. Assessing autonomic dysfunction in early diabetic neuropathy: the Survey of Autonomic Symptoms. Neurology 2011; 76(12) 1099-105.

[43] Thomas PK. Classification, differential diagnosis, and staging of diabetic peripheral neuropathy. Diabetes 1997; 46(2) S54-7.

[44] Rathur HM, Boulton AJ. Recent advances in the diagnosis and management of diabetic neuropathy. J Bone Joint Surg Br 2005; 87(12) 1605-10.

[45] Onde ME, Ozge A, Senol MG, Togrol E, Ozdag F, Saracoglu M, Misirli H. The sensitivity of clinical diagnostic methods in the diagnosis of diabetic neuropathy. J Int Med Res 2008; 36(1) 63-70.

[46] Freeman R. Not all neuropathy in diabetes is of diabetic etiology: differential diagnosis of diabetic neuropathy. Curr Diab Rep 9(6) 423-31.

[47] Shehab DK, Al-Jarallah KF, Abraham M, Mojiminiyi OA, Al-Mohamedy H, Abdella NA. Back to basics: ankle reflex in the evaluation of peripheral neuropathy in type 2 diabetes mellitus QJM 2012; 105(4) 315-20.

[48] van der Naalt J, Fidler V, Oosterhuis HJ. Vibration perception threshold, complaints and sensory examination in diabetic patients. Acta Neurol Scand 1991; 83(5) 297-300.

[49] van Deursen RW, Sanchez MM, Derr JA, Becker MB, Ulbrecht JS, Cavanagh PR. Vibration perception threshold testing in patients with diabetic neuropathy: ceiling effects and reliability. Diabet Med 2001; 18(6) 469-75.

[50] Manivannan M, Periyasamy R, Narayanamurthy VB. Vibration perception threshold and the law of mobility in diabetic mellitus patients. Prim Care Diabetes 2009; 3(1) 17-21.

[51] Baba M, Ozaki I. Electrophysiological changes in diabetic neuropathy: from subclinical alterations to disabling abnormalities. Arch Physiol Biochem 2001; 109(3) 234-40.

[52] Vinik AI, Kong X, Megerian JT, Gozani SN. Diabetic nerve conduction abnormalities in the primary care setting. Diabetes Technol Ther 2006; 8(6) 654-62.

[53] Kong X, Lesser EA, Potts FA, Gozani SN. Utilization of nerve conduction studies for the diagnosis of polyneuropathy in patients with diabetes: a retrospective analysis of a large patient series. J Diabetes Sci Technol 2008; 2(2) 268-74.

[54] Bourcier ME, Ullal J, Parson HK, Dublin CB, Witherspoon CA, Ward SA, Vinik AI. Diabetic peripheral neuropathy: how reliable is a homemade 1-g monofilament for screening? J Fam Pract 2006; 55(6) 505-8.

[55] Perkins BA, Orszag A, Ngo M, Ng E, New P, Bril V. Prediction of incident diabetic neuropathy using the monofilament examination: a 4-year prospective study. Diabetes Care 2010; 33(7) 1549-54.

[56] AstrupAS, Tarnow L, Rossing P, Hansen BV, Hilsted J, Parving HH. Cardiac autonomic neuropathy predicts cardiovascular morbidity and mortality in type 1 diabetic patients with diabetic nephropathy. Diabetes Care 2006; 29(2) 334-9.

[57] Callaghan BC, Cheng HT, Stables CL, Smith AL, Feldman EL. Diabetic neuropathy: clinical manifestations and current treatments. Lancet Neurol 2012; 11(6) 521-34.

[58] The Diabetes Control and Complications Trial Research Group. The effect of intensive treatment of diabetes on the development and progression of long-term complications in insulin-dependent diabetes mellitus. N Engl J Med 1993; 329(14) (Setember), pp. 977-86.

[59] Hotta N, Akanuma Y, Kawamori R, Matsuoka K, Oka Y, Shichiri M, Toyota T, Nakashima M, Yoshimura I, Sakamoto N, Shigeta Y. Long-term clinical effects of epalrestat, an aldose reductase inhibitor, on diabetic peripheral neuropathy: the 3-year, multicenter, comparative Aldose Reductase Inhibitor-Diabetes Complications Trial. Diabetes Care 2006; 29(7) 1538-44.

[60] Matsuoka K, Sakamoto N, Akanuma Y, Hotta N, Shichiri M, Toyota T, Oka Y, Kawamori R, Shigeta Y; ADCT Study Group. A long-term effect of epalrestat on motor conduction velocity of diabetic patients: ARI-Diabetes Complications Trial (ADCT). Diabetes Res Clin Pract 2007; 77(1) S263-8.

[61] Ramirez MA, Borja NL. Epalrestat: an aldose reductase inhibitor for the treatment of diabetic neuropathy. Pharmacotherapy 2008; 28(5) 646-55.

[62] Ziegler D, Nowak H, Kempler P, Vargha P, Low PA. Treatment of symptomatic diabetic polyneuropathy with the antioxidant alpha-lipoic acid: a meta-analysis. Diabet Med 2004; 21(2) 114-21.

[63] Papanas N, Maltezos E. α-Lipoic acid, diabetic neuropathy, and Nathan's prophecy. Angiology 2012; 63(2) 81-3.

[64] Kazakos KA, Sarafidis PA, Yovos JG. The impact of diabetic autonomic neuropathy on the incretin effect. Med Sci Monit 2008; 14(4) 213-20.

[65] Panchapakesan U, Mather A, Pollock C. Role of GLP-1 and DPP-4 in diabetic nephropathy and cardiovascular disease. Clin Sci (Lond) 2013; 124(1) 17-26.

[66] Drucker DJ, Sherman SI, Bergenstal RM, Buse, JB. The safety of incretin-based therapies--review of the scientific evidence. J Clin Endocrinol Metab 2011; 96(7) 2027-31.

[67] Hölscher C. Potential role of glucagon-like peptide-1 (GLP-1) in neuroprotection. CNS Drugs 2012; 26(10) 871-82.

[68] Isner JM, Ropper A, Hirst K. VEGF gene transfer for diabetic neuropathy. Hum Gene Ther 2001; 12(12) 1593-4.

[69] Hovaguimian A, Gibbons CH. Clinical Approach to the Treatment of Painful Diabetic Neuropathy. Ther Adv Endocrinol Metab 2011; 2(1) 27-38.

[70] Bril V, England J, Franklin GM, Backonja M, Cohen J, Del Toro D, Feldman E, Iverson DJ, Perkins B, Russell JW, Zochodne D; American Academy of Neurology; American Association of Neuromuscular and Electrodiagnostic Medicine; American Academy of Physical Medicine and Rehabilitation. (2011) Neurology 2011; 76(20) 1758-65.

[71] Vinik AI, Erbas T. Recognizing and treating diabetic autonomic neuropathy. Cleve Clin J Med 2001; 68(11) 928-30, 932, 934-44.